Acknowledgements

The author gratefully acknowledges the assistance of Wimpey Laboratories Limited in providing information, and the generous help of many members of the staff who gave advice and encouragement. Thanks are also due to R.N. Crookes, formerly of Wimpey Laboratories, for his contribution on the corrosion of reinforcing steel in concrete and for Fig. 4. Building Research Establishment is thanked for permission to use Figs 3 and 7, and Al Futtaim-Wimpey Laboratory Division, Dubai, is thanked for Fig. 5.

CONSTRUCTION LIBRARY

Liverpool Polytechnic
Clarence Street
Liverpool L3 5TN
Tel. (051) 207 3581 X 2634

LIVERPOOL POLYTECHNIC LIBRARY

3 1111 00193 0344

Eglinton, M.S
Concrete and its chemical behaviour
MP ABE 620.136 EGL 1987

Concrete and its chemical behaviour

M. S. EGLINTON BSc

🕇🕇 Thomas Telford, London

Published by Thomas Telford Ltd, Telford House, PO Box 101,
26–34 Old Street, London EC1P 1JH, England

First published 1987

British Library Cataloguing in Publication Data
Eglinton, M. S.
Concrete and its chemical behaviour
1. Concrete – chemistry
I. Title
666'-893   TP882.3

ISBN: 0 7277 0372 2

© M.S. Eglinton, 1987

All rights, including translation, reserved. Except for fair copying, no part of this publication may be reproduced, stored in a retrieval system or transmitted in any form or by any means, electronic, mechanical, photocopying, recording or otherwise, without the prior written permission of the publisher. Requests should be directed to the Publications Manager at the above address.

Set in Compugraphic Baskerville in 11/12 pt
by MHL Typesetting Ltd, Coventry

Printed and bound in Great Britain
by Redwood Burn Ltd, Trowbridge, Wiltshire

Preface

This book deals with the chemical properties of concrete and the materials from which it is made. The physical properties of the materials and methods for designing and making concrete are outside its scope, although resistance to chemical attack depends largely on the quality of the concrete, and in particular on its having a low degree of permeability to the ingress of aggressive solutions. The general assumption is made that concrete contains both fine and coarse fractions of dense aggregates and is used for most structural purposes. Cement-based materials used in building such as mortars, screeds and rendering coats are not included. Mention is made of lightweight aggregates, but lightweight aerated autoclaved concrete, foamed concrete and no-fines concrete, which are not of general application, are excluded.

Attention is given to problems in the use of concrete, to the conditions under which it might deteriorate, to the suitability of materials to perform well in aggressive conditions, and to precautions against chemical attack. The literature on these subjects is scattered and often not readily accessible, and it is hoped that the information given here will be of use to those concerned with concrete in civil engineering construction or will provide guidance on where further information may be sought. The author has drawn on Building Research Establishment and Cement and Concrete Association publications and on British and American standard specifications. Reference is also made to early work on some of the problems of current concern.

The main emphasis is on concrete made with Portland cements, although pozzolanic cements, the various types of cement made with blastfurnace slag, high alumina cement and the cement replacement materials – pulverised fuel ash and ground granulated blastfurnace slag – are discussed. The chemical compositions of cementitious materials and the compounds formed when they hydrate are outlined to give an understanding of their influence on the deterioration of concrete and the durability of concrete. Concrete aggregates, their chemical properties and their reactions in concrete form the subject of one chapter. The chemical agents which may be present in aggregates and which may cause damage to concrete are discussed, and the levels at which they may be tolerated in either the aggregate or the concrete are given. Causes of minor damage are mentioned.

Attention is given to the important subject of chemical attack on concrete from external sources. The use of concrete in soils containing sulphates, in sea water, in acidic waters and in other situations where attack may occur is a matter of concern. Decisions must be made on the risks of deleterious reactions taking place, on the types of cement most suitable for particular situations and on other protective measures that may be required. The natural aggressive agents are discussed in greater detail than those encountered in industrial situations, although severe problems may arise in construction on filled ground where chemical wastes might have been dumped, or in the redevelopment of sites formerly used for chemical manufacturing processes. These can only be dealt with on an individual basis, but the effects of groups of chemical substances are outlined. The protection of concrete used in the construction of manufacturing plants where it may be exposed to aggressive chemicals is a matter for specialist advice and is not discussed here.

Diagnostic features of various types of chemical attack on concrete are described and appropriate chemical tests to confirm the diagnoses are indicated. Standard specification and other test methods are discussed only in broad terms, with attention directed to their value and significance, rather than to their execution. Pure chemistry has been kept to a minimum by the omission of formulae and equations, and it is hoped that this approach will make the chemical reactions that can occur in concrete the more readily understood.

MSE
1987

Contents

1. Basic principles   1
   Definition of concrete, general durability, outline of factors governing quality, chemical resistance

2. Cements   4
   Hydraulic properties, Portland cements, blastfurnace slag cements, cements containing pozzolanas, high alumina cement

3. Concreting aggregates   21
   Natural aggregates containing sulphates and chlorides, aggregate reactions with cement alkalis, iron minerals in aggregates, the action of organic matter in natural aggregates; the chemical properties of dense and lightweight artificial aggregates

4. Other concreting materials   45
   Water for mixing and curing concrete, accelerating, retarding and water-reducing admixtures, superplasticising admixtures, air-entraining agents, integral waterproofers; polymer cement concrete

5. Chemical attack on concrete from external sources   53
   Action of sulphates, chlorides, sea water, soft waters, aggressive carbon dioxide, acidic and alkaline solutions, salts, sugars, fats and oils; action of gases; biological causes of attack

6. Examination of chemically attacked concrete 78
   Visual examination, patterns of cracking, seepages, staining, softening, spalling, expansive movements; sampling; laboratory and field testing techniques

7. Specifications for chemical testing 92
   Chemical tests on soils, ground waters and fills; analysis of cements and other cementitious materials; testing aggregates for sulphate, chloride, shell contents and for alkali–silica reactivity; analysis of hardened concrete

8. Protection of concrete against chemical attack 105
   Protection against acidic waters, sulphates, chlorides, sea water, alkali–silica reactions; selection of resistant materials; use of coatings and linings

Appendix 1. Atomic and molecular weights and molecular volumes 118

Appendix 2. Abbreviated formulae used in cement chemistry 119

References 119

Index 131

1
Basic principles

The term concrete applies to any artificial conglomerate in which particulate matter is bound into a solid mass, but for the purposes of this book is taken to mean only material in which the particulate matter, or aggregate, is bound with a hydraulic cement and water.

Concrete is one of the most versatile and widely used of all construction materials, and when it is properly made has a high degree of durability under normal conditions of exposure, as shown by the vast number of structures in which it has been used successfully over the past 100 years or so. It is, however, not immune to the action of many chemical substances, whether introduced with the materials used or entering from external sources. The extent to which concrete is able to withstand chemical attack depends primarily on its quality, a property which is not necessarily related to its strength, although many of the factors which determine quality also affect strength. Good quality stems from the design of the mix, the accurate proportioning of its constituents and adequate compaction after placing. Low permeability is a vital factor in the resistance of concrete to attack from external sources. Some types of cement are more resistant to certain chemical agents than are others, and cement replacement materials may offer a greater degree of resistance. The chemical nature of the aggregates used has a considerable influence on the durability of the concrete, as discussed in a later chapter.

It is not within the scope of this book to discuss in detail methods of producing concrete, but much information is available elsewhere. Mix design, making, transporting and placing, and the physical and mechanical properties required to ensure satisfactory performance under specific conditions of use are covered by standard specifications, codes of practice and other publications.[1-3] The practical aspects of concrete-making are the subjects of a number of authoritative texts, for example, references 4 – 6. All the same, it may be helpful to outline here the principles which govern the quality of concrete and hence the resistance it offers to chemical attack. The controlling factor is to be found in its name. The noun concrete is derived from a Latin verb meaning to grow together, and the more compactly the constituents of which concrete is composed can be made to grow together, in other words, the greater the denseness that can be achieved in its production, the more durable will it be.

The basic constituents of most structural concrete are a coarse aggregate, graded from 5 mm in particle size up to a maximum generally not exceeding 40 mm, a fine aggregate, or sand, graded from 5 mm down to 75 μm, cement and water. The coarse aggregate comprises the bulk of the mix, with the sand and cement paste filling the spaces between the particles. It is important therefore that its grading is such that it packs together leaving as few voids as possible. The mix should include sufficient fine aggregate to fill the voids, and this also should be graded over a range of particle sizes. The shape of the coarse aggregate particles, and, to a lesser extent, of the fine aggregate, affects the efficiency of their packing together. Particles which are equidimensional in shape pack more compactly than those which are angular, flaky or elongated. Methods for determining aggregate gradings and the particle shape of coarse aggregate are given in BS 812.[7] Limits are given in BS 882.[8] The tests and the practical applications of their results have been discussed in a recent publication,[9] one chapter of which deals comprehensively with the physical and chemical properties of aggregates for use in concrete and their effects on concrete quality.

The quality of concrete is chiefly governed by the impermeability of the cement and fine aggregate part of the mix. The coarse aggregate plays little part, provided that it is not so porous as to provide a channel for the ingress of aggressive solutions. The chief factor controlling the permeability of the fine fraction is the amount of water used in relation to the cement content of the mix, that is, the water/cement ratio. The amount of water required in any one mix depends on a number of fac-

Basic principles

tors, such as the surface area and absorbent properties of the aggregate, but for the production of durable concrete it should be kept as low as possible. No more should be used in the mix than is necessary to ensure that all aggregate particles are completely covered by the cement paste, that the cement and fine aggregate mortar fills the voids between the coarse aggregate, and that the concrete is fluid enough to be placed and compacted while remaining coherent. These practical requirements mean that more water must be used in mixing than is needed for the hydration of the cement. The latter is about 20–25% for Portland cements, but about twice as much may be required to produce workable concrete. Any excess of water over the minimum requirement is detrimental, leading to an increase in the number of voids which will be left as the concrete dries, thereby increasing the porosity of the cement and fine aggregate mortar, rendering it more vulnerable to chemical attack, and also reducing its strength. Over-wet concrete is prone to segregate during placing, with an excess of coarse aggregate settling towards the bottom of the pour, while a weak layer composed of cement, fine sand and water, called laitance, forms at the surface. These segregated layers in concrete constitute planes of weakness at which chemical attack may begin. The use of plasticising admixtures to give the mix the workability required at a lower water/cement ratio can increase the chemical resistance of the concrete by reducing its permeability. A lower water/cement ratio will also increase the strength of the concrete, or allow a reduction in cement content for a given strength, but where chemical resistance is required the higher cement content should be maintained.

Another factor governing the strength and durability of concrete is its curing. This means preventing loss of water during the early stages of the hydration of the cement by providing the concrete with a moist environment at its critical early age, either by applying water to the surface or by preventing loss of water from the surface. A discussion of the methods and materials which may be used would be out of place here, but it must be emphasised that correct curing plays an important part in the production of good quality concrete.

2
Cements

INTRODUCTION TO HYDRAULIC CEMENTS

Hydraulic cements are finely ground materials which, when mixed with water, set and harden either in air or in water by forming hydrated compounds which increase in strength with age. Their principal constituents are compounds formed by reactions between calcium oxide, aluminium oxide and silica, which are present in the raw materials used in their manufacture. Hydraulic cements are sometimes described as calcareous cements, and although calcium oxide is not necessarily the preponderant constituent, its presence is essential for the formation of the aluminates and silicates which give the cements their binding powers. Materials such as gypsum plaster and lime were used from early times as cementing agents in masonry and concrete construction, but neither can be described as truly hydraulic. Plaster sets by direct recrystallisation on the addition of water, and lime hardens by atmospheric carbonation.

The materials used in construction in early civilisations were often impure, and their cementitious properties may have been due to more than one hardening reaction. Lime produced by burning gypsiferous limestone might have contained some plaster, and a proportion of quick lime might have been present in plasters obtained from gypsum deposits interbedded with limestones, even though the calcining temperatures of the two materials are markedly different. The mortar used to joint the limestone blocks of the pyramids of Egypt, built between 4000 and 2000 BC, was made with an impure gypsum plaster and has been found to contain calcium carbonate as well as calcium sulphate. An important impurity in limestone would have been clay, but its significance in giving the calcined limes hydraulic properties appears not to have been recognised until the experiments made by Smeaton in the second half of the 18th century when he was commissioned to rebuild the Eddystone lighthouse. He found that the mortars which performed best under water were those made with limes containing a fairly high proportion of clayey material. Limes of this type, now termed hydraulic limes, can be considered as having a chemical composition which is between that of pure limes and Portland cements. They contain calcium silicates that have variable, but generally weak, hydraulic properties; but, unlike Portland cements, they also contain considerable free lime and therefore slake in water.

The first truly hydraulic cementing agent was the mixture of pure lime and volcanic ash which was known to both the Romans and the Greeks to have the property of hardening under water. The Greeks obtained their ash from the island of Santorini, where deposits of the earth are still worked. The Romans used tuffs found in Pozzuoli near Naples, from which the modern name of pozzolana for materials of similar properties is derived. Where natural volcanic earths were not available, as in Britain, the Roman builders substituted crushed fired clay tiles and bricks, which they found increased the hardness and durability of their lime mortars and gave them a degree of hydraulicity. Mortars made with volcanic earths and lime were used extensively by the Romans in coastal works, some of which remain in good condition. Despite Smeaton's discovery of the greater hydraulicity of mortars made with hydraulic lime, the use of pure lime – pozzolana mixes continued in popular use in the UK until a gradual appreciation of the advantages of more hydraulic materials, particularly for use under water and in marine exposure conditions, led in stages and through trials on many materials to the invention of Portland cement in the mid-19th century.

The cement industry grew rapidly throughout Europe and America, and within the next half century had become so well established that it was possible for standard specifications for the composition and performance of Portland cements to be published in Germany, the UK and the USA.[1] With few exceptions, all cements used today consist entirely of Portland cement or contain a proportion of it, as in blastfurnace slag and pozzolanic cements.

PORTLAND CEMENTS

Portland cements consist principally of compounds of calcium and silica with smaller amounts of aluminium and iron compounds. They are made from homogeneous, balanced mixtures of calcareous and argillaceous materials. Chalk or limestone is generally used as the calcareous component and clay or shale as the argillaceous, but such is the simplicity of their raw material requirement that Portland cements can be manufactured from any materials which will provide the required balance of compound-forming constituents. Calcareous clays, such as marls, have been found suitable, while the Portland cement industry in the USA was founded on the use of an argillaceous limestone, known as cement rock. The raw materials are sintered, that is, not completely melted, at a temperature which allows only a part to become liquid. The resulting product, the clinker, is therefore not homogeneous, but consists of a number of solid phases. The clinker is interground with a small proportion, about 3–5%, of gypsum or other forms of calcium sulphate to produce the finished cement.

The complex chemistry of Portland cement and its hydration have been the subjects of many detailed studies and are not yet fully resolved,[2] but an accepted simplification is that it consists of four main compounds – tricalcium silicate, dicalcium silicate, tricalcium aluminate and tetracalcium aluminoferrite. Minor constituents are present within these phases, but with the exception of magnesium oxide and the alkali metals, sodium and potassium, they are of no practical significance. The amounts in which the four main compounds are present are calculated from the chemical analysis of the cement using formulae proposed by Bogue. Latterly, modifications of the formulae, derived from the results of X-ray diffraction analyses and considered to be more precise, have been adopted, but the Bogue calculations are still included in British and American Standard specifications for Portland cements.

The compound composition of a cement and the manner in which each of its phases reacts with water govern its physical and chemical properties. Some knowledge of the hydration processes is therefore desirable for an understanding of the chemical reactions which can occur in concrete and how they may be modified or counteracted by altering the composition of the cement or by adding materials which have greater chemical resistance.

The calcium silicates, which together comprise 70–80% of Portland cements, contribute most to binding power and strength. Hydration of tricalcium silicate begins very early and continues at a steady rate and with a moderate evolution of heat up to about 28 days, when it is mainly complete. Virtually no hydration occurs later than one year. Its principal contribution is to the early strength of the cement. Dicalcium silicate hydrates much more slowly. The greater part of its hydration does not take place until after about 28 days, but it continues beyond one year. The contribution of dicalcium silicate to the strength of the cement is therefore at later ages. The slow rate of hydration also means that the heat of hydration is low. The overall hydration reaction of both silicates is the formation of a gel of calcium silicate hydrates, which is the main binding agent in the cement paste, and of crystalline calcium hydroxide. Calcium hydroxide is the compound most readily leached from Portland cement concrete when exposed to soft or acid waters. It also reacts with some sulphate solutions.

Tricalcium aluminate exhibits flash setting on hydration, that is, it sets almost instantaneously, with the evolution of a considerable amount of heat. In Portland cements the set is retarded by the calcium sulphate (gypsum or anhydrite) added to the clinker. The initial product of the reaction between tricalcium aluminate and calcium sulphate is a form of tricalcium sulphoaluminate containing three molecules of calcium sulphate and 31 molecules of water. This compound is generally known as ettringite, but is also described as high-sulphate tricalcium sulphoaluminate. It is insoluble in the calcium hydroxide present in the pore water of the hydrating cement, and forms a coating over the aluminate particles, delaying their continuing hydration. As crystals of ettringite develop, the coating becomes disrupted, exposing fresh surfaces of the aluminate and allowing its hydration to proceed. Ettringite is formed during the first 24 hours of hydration of the cement, but within the next few days this is largely converted to a low sulphate form of tricalcium sulphoaluminate containing only one molecule of calcium sulphate and 10 molecules of water. Both the high and the low sulphate forms have been found to persist indefinitely in the hardened cement paste.[3] Only a small amount of gypsum, up to about 5%, can be added to the clinker without causing unsoundness in the cement, and in ordinary Portland cements a large proportion of the tricalcium aluminate, which forms 5–12% of the cement, remains unaltered. Tricalcium aluminate influences the early set of the cement and possibly contributes to its early strength, but has little effect on its ultimate physical properties. It is the component of hardened Portland cement pastes which is most vulnerable to attack by sulphate solutions.

Tetracalcium aluminoferrite reacts at a slower rate than tricalcium aluminate and contributes little to the strength of the cement at any age. It is thought to form both high and low-sulphate forms of tetracalcium aluminoferrites during its hydration, in the same manner as for tricalcium aluminate. Very little heat is evolved during its hydration.

The differences in behaviour of the four main compounds make it possible for a range of Portland cements to be produced, each having its own special properties. These are mainly physical, and affect such properties as the rate of hardening, of strength development and of evolution of heat on hydration. The different types of cement manufactured to British standard specifications are described in Building Research Establishment Digest 237.[4] The most important chemical difference is in the resistance offered to sulphate attack. Since tricalcium aluminate is the compound most affected by sulphates, limiting its content in the cement results in increased resistance. This is accomplished by adjusting the raw material mix so that a greater pro-

Table 1. Permitted levels of constituents in cements to BS 12[7] and BS 4027[5]

Constituent	Percentage by weight
Loss on ignition	3.0 (4.0 in tropical climates)
Insoluble residue	1.5
Magnesium oxide	4.0
Sulphate	2.5 (if the tricalcium aluminate content is less than 7%)
	3.0 (if the tricalcium aluminate content is more than 7%)

portion of tetracalcium aluminoferrite forms in the clinker at the expense of the aluminate phase. Ordinary Portland cements contain about 8–11% of tricalcium aluminate. Sulphate-resisting Portland cements contain between 3 and 5%. The maximum allowed in cements manufactured to BS 4027[5] is 3.5%. A maximum of 4% is allowed in ASTM C150–81[6] for type V cements, intended for use where a high degree of sulphate resistance is required. In cements required to withstand only moderate concentrations, ATM type II cement, the maximum permitted content of tricalcium aluminate is 8%. Only the tricalcium aluminate content is limited in BS 4027, but a number of other national specifications also place limits on the contents of tetracalcium aluminoferrite or of the combined ferrite and aluminate phases.

Concrete made with sulphate-resisting Portland cement is not immune to attack by very high concentrations of naturally occurring sulphates in soils or ground waters. In the presence of sulphates other than those which occur naturally, for example, ammonium sulphate which is highly aggressive, the performance of sulphate-resisting Portland cement differs little from that of ordinary Portland cement. It is considered to be slightly more resistant to the action of weak acids than ordinary Portland cement, which is not recommended for use in concrete to be exposed to pH values of less than about 6, but limiting values have not been established. The effect of chlorides on sulphate-resisting Portland cement is discussed in Chapter 3.

British Standard specifications BS 12[7] and BS 4027[5] place the maximum limits shown in Table 1 on controlled constituents of Portland cements.

The range of maxima allowed in other national standard specifications is shown in Table 2. In some countries, the maximum sulphate allowed is based on a tricalcium aluminate content in the cement above or below 8%.

*Limits of between 0.66 and 1.02 are stipulated in the British specifications and in a number of others for what is called the lime saturation factor. This is calculated from the chemical analysis of the cement and represents the balance between the calcium oxide and the silica and aluminium and iron oxides which are the main compound-forming components of the clinker. The reason for controlling magnesium oxide contents is perhaps less obvious. Magnesium oxide does not form any particular phase in the cement, but enters the main compounds by solution in the liquid which forms during clinkering. However, only a small amount can be dissolved and any excess that may have been present in the feed materials crystallises as an oxide, the mineral periclase. Although periclase can eventually hydrate to form magnesium hydroxide, it is at a very slow rate. Its hydration is accompanied by a volume increase, which in a hardened cement paste can result in cracking of the concrete. The rate at which periclase hydrates is dependent on its crystal size, which in turn is governed by the rate at which the clinker cools. In countries where the available raw materials are such that it is not realistic to specify a very low magnesium oxide content in the cement, the likelihood of unsoundness due to the presence of periclase is generally assessed by specified tests for the measurement of expansion under conditions of accelerated hydration.

A question which often arises, particularly in overseas work, is whether cements which conform to the chemical requirements of the countries in which they are produced, although these may differ from requirements in the UK, are acceptable when job specifications require materials to be to British Standards. The answer is that in the case of Portland cements they generally are acceptable. The differences are fairly minor, and, in some aspects, without significance. For example, the insoluble residue represents non-cementitious, or inert, material, and an increase of $1-2\%$ in its content in the cement is well within the

Table 2. Level of constituents in Portland cements given in international standards

Constituent	Percentage by weight
Loss on ignition	3–5
Insoluble residue	0.75–3
Magnesium oxide	2.5–6.5
Sulphate	2.3–3.5 (if the tricalcium aluminate content is less than 7%)
	3–4 (if the tricalcium aluminate content is more than 7%)

accuracy of batching the concrete. Again, although in Tables 1 and 2 the sulphate contents are given in relation to the tricalcium aluminate content of the cement, many national specifications have a single limiting value for sulphate, irrespective of the amount of aluminate present.

BLASTFURNACE SLAG CEMENTS

Blastfurnace slag is produced simultaneously with iron in the smelting of iron ore. It results from the combination in a liquid state of silica and aluminium compounds present in the ore with calcium and magnesium oxides formed by the decomposition in the furnace of calcium and magnesium carbonate rocks used as fluxes. Blastfurnace slag consists essentially of the same oxides as Portland cement clinker, namely calcium and aluminium oxides and silica, but in different proportions. The mineral phases formed are therefore different from those of Portland cement. The major phases in slag are calcium silicates and calcium aluminium silicates. When magnesium limestones or dolomites have been used in the flux, the compound assemblages include magnesium silicates and aluminates and silicates containing both calcium and magnesium. Iron oxide is present only in low concentrations, typically less than 1%, and does not form a ferrite phase as in Portland cements. A function of the molten slag is to remove sulphur from the iron in the form of calcium sulphide, which is present in all slags. Less commonly the sulphur may from iron or manganese sulphides.

Since slag is a by-product, its chemical composition varies, between quite wide limits, as illustrated in Table 3 by figures compiled from the analyses of samples taken from different sources. Blastfurnace slag is only hydraulic when cooled under such conditions that it solidifies as a glass. When allowed to crystallise, it has no cementitious properties. To solidify as a glass the molten slag must be cooled very rapidly from as high a temperature as possible. This is done by quenching it with water as it leaves the furnace. The resulting product, termed granulated slag, is a light frothy mass, ideally with a high proportion of glass beads. Not all slags can be successfully granulated for use in cements. Chemical composition influences both the glass-forming properties and the hydraulicity of the slag, the chief criterion being the relative proportions of the calcium, magnesium and aluminium oxides to the silica. Blastfurnace slag develops its cementitious properties far too slowly to be of practical use unless its hydration is activated by the addition of calcium compounds. The earliest used was hydrated lime (calcium hydroxide). This has been replaced by Portland cement, which acts in the same way through the release of calcium hydroxide on hydration, and has the advantage of giving early strength. Calcium sulphate is used as the main activator in one class of slag cements.

Portland–blastfurnace slag cements are produced either by intimately mixing finely ground granulated slag with Portland cement or by intergrinding the slag with Portland clinker. In the latter case, an appropriate amount of gypsum or anhydrite is also interground to regulate the set of the Portland cement. Portland–slag cements are widely produced throughout the world, although not to any great extent in the UK. Two types are defined in British Standards – Portland–blastfurnace cement is governed by BS 146[8] and low heat Portland–blastfurnace slag cement by BS 4246.[9] The former contains a maximum slag content of 65%, while the latter may have a slag content of between 50 and 90%. In the specifications of other countries slag contents range from 10 to 95%.

The chemical constituents controlled by specification are loss on ignition, insoluble matter, magnesium oxide, sulphate and total sulphur contents. The maximum amounts permitted in the British standards are given in Table 4. These limits are generally

Table 3. Composition of blastfurnace slags

Constituent	Percentage by weight
Silica (SiO_2)	30–35
Aluminium oxide (Al_2O_3)	13–23
Iron oxide (Fe_2O_3)	0.3–1
Calcium oxide (CaO)	34–43
Magnesium oxide (MgO)	11–22
Total sulphur (S)	1–2.5
Sulphate (SO_3)	0.1–0.5
Titanium dioxide (TiO_2)	0.5–1
Manganese oxide (MnO)	0.3–1.5

in line with those adopted in other national standards. It may seem surprising that such high contents of magnesium oxide are permitted in Portland–blastfurnace slag cements, while being rigorously controlled at a much lower level in Portland cements. The reason is that in slag cements the magnesium oxide is contributed mainly by the slag, in which it is fully combined in silicate and aluminate compounds. The free oxide, periclase, which if present in Portland cements can cause expansion on hydration, has not been observed in blastfurnace slags.

The setting of Portland–blastfurnace slag cements takes place in two stages. On mixing with water, the Portland cement hydrates in its normal manner, liberating calcium hydroxide from its di- and tricalcium silicates. The calcium hydroxide then gradually reacts with the slag minerals to form other hydrated compounds. Blastfurnace slag does not contain tricalcium silicate, which is the phase contributing most to the early strength of Portland cements, and therefore slag cements develop their strength at a slower rate. Nor does slag contain tricalcium aluminate, which is the phase of Portland cement having the greatest heat of hydration, and which also enters into expansive reactions with sulphates. The use of Portland–blastfurnace slag cements is therefore of benefit in reducing temperature rise in large masses of concrete, and in providing increased sulphate resistance. The general properties of the cements can be modified by the use of rapid hardening, high early strength or sulphate-resisting Portland cements or clinkers.

Ground granulated blastfurnace slag may also be added to Portland cement at the mixer to provide protection against the action of sulphates. The mixtures are not strictly considered to come within the definition of Portland–blastfurnace slag cements and are not governed by standard specifications. To be effective when used in this way the slag is required to have a suitable chemical composition and a high glass content. A British Standard is being prepared.

A slag cement in which the activator is calcium sulphate is known in the UK as supersulphated cement and elsewhere in Europe as ciment sursulfaté or ciment métallurgique sursulfaté. The cement called ciment permétallurgique is not of this type, but is a Portland–blastfurnace slag cement containing a high proportion of slag. The various names often give rise to confusion regarding their composition when these cements are encountered in overseas work.

Supersulphated cements contain between 80 and 85% of granulated blastfurnace slag, 10–15% of anhydrous calcium sulphate in the form of anhydrite or burnt gypsum and about 5% of Portland cement. The Portland cement is not included to influence the rate of setting or hardening of the cement. Its function is to provide calcium hydroxide which is essential for the hydration reactions which take place between the slag and the calcium sulphate. Hydrated lime can be used in its place. The mode in which supersulphated cements set and harden is different from that of Portland–blastfurnace cements. The main products formed on hydration are ettringite and a calcium silicate phase of uncertain composition, but probably containing the magnesium oxide and part of the aluminium oxide present in the slag. Hydrated aluminium oxide may also be present as a separate phase. The calcium hydroxide contributed by the Portland cement or lime included in the formulation enters the main phases and is absent in the set cement. Ettringite, which comprises about 40% of the set cement, forms rapidly and is responsible for early strength development, the other compounds contributing at a later stage. The combination of aluminates as ettringite and the absence of free calcium hydroxide give supersulphated cements a high degree of resistance to the action of most sulphate solutions and to acids down to a pH value of about 3.5.

Table 4. Permitted levels of constituents in cements to BS 146[8] and BS 4246[9]

Constituent	Percentage by weight	
	BS 146	BS 4246
Loss on ignition	3.0	
Insoluble residue	1.5	1.5
Magnesium oxide (MgO)	7.0	9.0
Sulphate (SO_3)	3.0	3.0
Total sulphur (S)	1.5	2.0

Although supersulphated cement is not currently manufactured in the UK, requirements for its chemical composition and physical properties are detailed in BS 4248.[10] Its use is recommended in BS 8110[11] when concrete is to be exposed to high levels of sulphates in soils or ground waters.

CEMENTS CONTAINING POZZOLANAS

Pozzolanas are siliceous or alumino-silliceous materials which, while in themselves possessing little or no cementitious properties, can react with lime in the presence of water and at ordinary temperatures to form stable hydrated compounds which are cementitious. It has already been mentioned that lime–pozzolana mixes were used for structural purposes before the development of Portland cement. Modern pozzolanic cements all contain Portland cement which on hydration provides the calcium hydroxide (lime) necessary for the formation of cementitious calcium silicates and aluminates and which also gives the concrete early strength. The pozzolanas originally used were volcanic earths, and although these are still used in countries where they occur, a greater number of cements produced throughout the world incorporate other materials which have pozzolanic properties either in their natural state or after activation by heat treatment. These include diatomaceous earths, cherts, shales and clays. By-products, such as spent oil shales and pulverised fuel ash, are also used. Blastfurnace slag, which is excluded from the International Standards Organisation[12] definition of pozzolana, is, in a number of countries, included in pozzolanic cements in combination with either natural pozzolana or pulverised fuel ash.

Essential requirements for pozzolanicity are that the material should be siliceous, glassy and finely divided. Chemical composition alone does not indicate activity. Thus, the volcanic earths consist chiefly of aluminium silicates, while the diatomaceous earths, which are composed of the skeletal remains of minute aquatic organisms, consist of amorphous, opaline silica. The activity of some cherts and shales is also due to the presence of opal. Clays are essentially aluminium silicates, but are inactive unless calcined. Pulverised fuel ash, also known as fly ash, is a residue from the combustion of powdered coal in the furnaces of power station plants. Being very fine and light, it is carried away from the grate by the flue gases and is separated from them by either mechanical or electrostatic means. It leaves the furnace in the form of fused particles, and cools to spherical glass beads, composed mainly of aluminium silicates. Its chemical composition is similar to that of a burnt clay, but varies widely depending on the coal burnt and the operating conditions of the furnace.

The international standards define two types of cement containing pozzolanas — Portland-pozzolana cement and pozzolanic cement, both of which consist of homogeneous mixtures of Portland cement and finely ground pozzolana, with the possible addition of calcium sulphate. The term Portland-pozzolana refers to cements which contain less than 20% of pozzolana and which do not necessarily satisfy the chemical tests for pozzolanicity given in BS 4550.[13] Cements termed pozzolanic contain up to 40% of pozzolana and do satisfy the pozzolanicity tests.

Pulverised fuel ash is the only pozzolana used in the UK, mainly as an addition at the mixer. Requirements for its chemical and physical properties are given in BS 3892.[14] There are also two specifications for cement containing it. These are BS 6588[15] for Portland pulverised fuel ash cement, and BS 6110[16] for pozzolanic cement with pulverised fuel ash as pozzolana. The respective contents of ash are 15–35% and 35–40%. In line with the international definitions, the latter cement is required to satisfy the chemical test for pozzolanicity while the former need not do so. The two cement specifications and that for pulverised fuel ash give limits of 4.0% for magnesium oxide content. Sulphur compounds are restricted to 3.0% in the cements and to 2.5% in the ash. Maximum permitted losses on ignition are given as 4.0% in BS 6588, 4.5% in BS 6110 and 7.0% in BS 3892. The ignition loss of the pulverised fuel ash is due to particles of unburnt fuel which accompany the ash in the gas stream. The potential pozzolanicity of the pulverised fuel ash is assessed, not by the chemical test, but by comparing the compressive strengths of mortar bars made with a combination of ash and Portland cement with those of bars made with Portland cement only. Although chemical composition alone is of little value in assessing pozzolanic activity, the

Cements

ASTM specification for pozzolanas[17] includes a requirement that the contents of silica, aluminium oxide and iron oxide should total not less than 70%.

Pozzolanic cements have a greater resistance than ordinary Portland cement to chemical attack by sea water, sulphates and soft or slightly acidic waters. The mechanism of their increased chemical resistance has not been fully elucidated, although immobilisation of the calcium hydroxide of the Portland cement plays a large part in reducing leaching of the concrete by soft or acidic waters. The resistance to sulphate attack is less clear, since pozzolanic cements contain calcium aluminates derived both from the Portland cement and from reaction between calcium hydroxide in the cement and aluminium oxide in the pozzolana, and these might be expected to react expansively with sulphate solutions. Theories have been proposed to explain why this does not happen. These are that the inclusion of pozzolana increases the impermeability of the concrete, rendering it less vulnerable to ingress of aggressive solutions; that as the concentration of calcium hydroxide in the pore water is reduced the calcium aluminates become more soluble and ettringite is formed in an aqueous, not solid, phase; and that because of the increased amount of silica contributed by the pozzolana the hydrated cement contains more calcium silicate gel than Portland cement pastes and the gel forms a protective film on the aluminates. The last theory is supported by the work of Rio and Celani[18] who found that sulphate resistance was related to the ratio of reactive silica to reactive aluminium oxide in the set cement, and that the resistance was increased by the use of pozzolanas rich in reactive silica in combination with cements of low aluminium oxide content. However, it is specifically stated in BS 8110 that pulverised fuel ash should not be used in combination with sulphate-resisting Portland cement in concrete which may be exposed to sulphate attack.

HIGH ALUMINA CEMENT

High alumina cement has had a chequered history. It was developed in France at the beginning of the 20th century after the widespread failure of Portland cement concrete placed in gypsiferous strata had indicated the need for a sulphate-resisting cement. At that time the advantage of a low tricalcium aluminate content in conferring sulphate resistance to Portland cement had not been realised. High alumina cement proved to have excellent sulphate resistance and other valuable properties, in particular, a high early strength, which later led to its extensive use in precast concrete manufacture. However, early lack of knowledge of its sensitivity to certain conditions of exposure brought it some disrepute. Within about 30 years serious faults had been found in a number of structures, and its use, except by licence, was banned in France in 1943 and later in Germany. In 1973, the collapse of precast, prestressed roof beams over a swimming pool in a London school led to its exclusion for structural purposes from British Standard codes of practice and deletion from the Building Research Establishment's recommendations for concrete to be placed in sulphate-bearing soils and ground waters.

High alumina cement is of different chemical composition from Portland cements. It is made by the complete or partial fusion of materials containing mainly calcium and aluminium oxides with smaller proportions of iron and other oxides. British Standard specification 915[19] requires the mix to be fused to a completely molten state. The raw materials are normally limestone and bauxite, which is an ore of aluminium containing about 50–55% of aluminium oxide and about 25% of iron oxides. The cement contains roughly equal amounts of aluminium oxide and calcium oxide, from 36 to 40% of each, and about 16% of iron oxides. The silica content generally does not exceed 5%, and sulphate and alkali metal contents are negligible. The requirements of BS 915 are that the aluminium oxide content should be not less than 32%, and that the ratio of aluminium oxide to calcium oxide should be between 0.85 and 1.30. The cement therefore consists mainly of calcium aluminate, together with mixed ferrites formed from the iron oxides. Only a small amount of calcium silicate is present.

In simplified terms, the hydration of high alumina cement can be considered to be the formation of monocalcium aluminate containing 10 molecules of water. This is metastable and in time converts to a stable tricalcium aluminate containing six molecules of water. Hydrated aluminium oxide also forms and the overall reaction results in the release of 18 molecules of free water. The molecular volumes of the products of conversion are less than those of the original hydration products, so conversion results in increased porosity of the cement paste with consequent loss in strength of the concrete. The conversion process is very slow at low temperatures, below 15°C, but is accelerated at high temperatures, particularly under humid conditions. Lea[20] gives the time for half conversion of neat cement specimens (water/cement ratio 0.26) when stored at different temperatures as one week at 50°C, 100 days at 40°C, and an estimated 20 years at 25°C. A concrete cube stored in water at a constant 18°C showed that the cement had half converted in 27 years, whereas no conversion had occurred after 30 years in the concrete of a pile taken from the sea bed, where the sea-water temperature never exceeded 18°C.[21] Loss of strength is also a function of the rate of conversion, and a cement which has converted slowly at a low temperature will lose considerably less strength than one which has converted rapidly at a high temperature.

The loss of strength on conversion can be offset by using less water in mixing the concrete than is required for the complete initial hydration of all the cement used. In theory, full hydration requires a water/cement ratio of about 1.0. Lafuma[22] considered that the amount of water used in mixing should not be more than could be combined in tricalcium aluminate hexahydrate, which is the conversion product of the cement. This amount of water is equivalent to a water/cement ratio of 0.45, a level at which only about 40% of the cement would be hydrated initially. As the conversion reaction proceeds, water which is set free during the alteration of the decahydrate to the hexahydrate form becomes available to hydrate the anhydrous cement, and this continuing process of hydration helps to maintain the strength of the concrete.

In addition to having good resistance to the action of sulphates, high alumina cement is resistant to the effects of acid solutions down to a pH value of 4, and to leaching by soft waters. It is attacked by alkali solutions which promote conversion. An early case noted of damage to high alumina cement concrete occurred in poles set in blocks of Portland cement concrete buried below ground level, and was caused by capillary movement of alkalis from the Portland cement into the high alumina concrete. The chemical reactions, which have been termed alkaline hydrolysis, resulted in decomposition of the calcium aluminates into hydrated alumina gel, of little binding power. Lafuma warned that the same reactions could occur if the aggregates contained alkalis that could be released into solution, and recommended that granites, schists and micaceous sands should not be used in high alumina cement concrete in size fractions less than 0.5 mm.

The chemical resistance of high alumina cement is reduced after conversion, whether this occurs naturally with age or is accelerated by exposure to high temperatures. The porosity of the converted concrete is influential in allowing ready ingress of aggressive solutions, but the products of conversion are also more susceptible to chemical attack.

Despite its uncertain performance in structural concrete, high alumina cement gives good service in refractory concrete, and by suitable choice of aggregate can be used up to 1600°C. At high temperatures it acts as a ceramic not a hydraulic cement. On exposure to heat the hydraulic bond is lost as the combined water is driven off, and as the temperature increases the ceramic bond develops by solid reactions between the cement and the fine aggregate. Conversion is therefore not a factor in the refractory use of high alumina cement.

3
Concreting aggregates

NATURAL AGGREGATES

The UK is fortunate in having ample supplies of sound natural aggregates, which include limestone, sandstone, granite, basalt, dolerite, flint gravel and silica sand. This is not always the case in other parts of the world, especially in arid regions, where available aggregates may be contaminated with soluble salts, in particular sulphates and chlorides, both of which can react adversely in concrete. The presence of sulphates can lead to chemical attack on the aluminates of Portland cements, while chlorides can cause corrosion of reinforcing steel. Both reactions are expansive and result in cracking and disruption of the concrete. Reactions may also occur between certain mineral constituents of aggregates and the hydroxides of the alkali metals, sodium and potassium, present in the cement, resulting in cracking of the concrete.

Some rocks exhibit marked shrinkage on drying, and when used as concreting aggregate can cause cracking,[1] but this is a physical property of the rock and chemical reactions are not involved. Undesirable impurities, whose effects are also physical, include excessive numbers of shells in marine-dredged aggregates, mica in concreting sands, and soft materials such as clay lumps or coal, and limits have been given for their permissible contents.[2,3]

Minor defects, such as surface disfigurement caused by staining or popping, can result from chemical reactions between cement and aggregate constituents, but in general no structural damage is caused.

Sulphates in aggregates

The presence in aggregates of any type of soluble sulphate in other than limited amounts can cause chemical attack to take place in Portland cement concrete. Sulphates react with the hydrated aluminates in the cement to give high-sulphate tricalcium sulphoaluminate or ettringite. This occupies more than twice the molecular volume of the aluminate and its formation in hardened pastes is accompanied by expansive forces which can exceed the tensile strength of the concrete. Sulphate attack is more rapid and severe with cements of high tricalcium aluminate content, but even sulphate-resisting Portland cements are not immune to the effects of large amounts of sulphate in the concrete.

The most common naturally-occurring sulphate likely to be present in aggregates is calcium sulphate in the form of gypsum ($CaSO_4 \cdot 2H_2O$). In hot regions of the world where surface temperatures can exceed 40°C, the gypsum may be partially or completely dehydrated through the bassanite series of sulphates containing less than two molecules of water to anhydrite ($CaSO_4$). Barium sulphate, barytes, and strontium sulphate, celestite, are of such low solubility in water that they do not contribute to sulphate attack. Barytes is used as a heavy aggregate in concrete for nuclear shielding. Celestite, while of rarer occurrence than gypsum, is known to be present in small amounts in some aggregates, notably in the Middle East. Its solubility in acids is also low, and it goes undetected in the usual wet chemical tests for sulphate content. Deposits of the very soluble sulphates of magnesium (Epsom salt) and of sodium (Glauber's salt) occur naturally but usually not in formations which yield usable aggregates.

Damage to concrete caused by sulphates in natural aggregates is not a problem in the UK, where the sulphates occur chiefly in clay strata. It might be thought that, because of its high sulphate content, 2 g per litre as SO_3, residual sea water would be a source of contamination of marine-dredged sands and gravels, but that is not the case. The sea-won materials are composed principally of flint, which has a low water absorption, so any contamination is on the surface only. If it is assumed that aggregate stockpiles retain 10% of water after draining, even if unwashed, marine aggregates would not contain more than

0.02% of sulphate. Pit sands and gravels and crushed rock aggregates are naturally of low sulphate content. The formation of sulphates by the oxidation of iron pyrite is described later under Iron minerals in aggregates.

However, in other parts of the world, particularly in arid regions, aggregates are not necessarily free from sulphate contamination. Igneous rocks are low in sulphates, but carbonate rocks of sedimentary or evaporite origin may be heavily contaminated. Sands, whether from inland pits or from beach deposits, may also be heavily contaminated. Tests made over several years of sands taken from one pit in the Middle East showed that sulphate contents averaged 1.0% SO_3, with individual samples containing as much as 5.0%. Crushed carbonate rock aggregate from the same area had an average of 0.6% SO_3 within a range of 0.06–1.20% SO_3. In another survey, sand samples taken from above the water table along an 18 km stretch of beach had an average sulphate content of 0.6% SO_3. These values are not unusual for sulphate-bearing Middle East aggregates. Calcium sulphate has been found to be most commonly present in both coarse and fine aggregates, but sands may contain the more soluble magnesium, sodium and potassium sulphates. These can amount to 30% of the salts present in saline ground waters and 60% of the total sulphates present. In such conditions, sands taken from below the water table may contain appreciable concentrations. If the ground water level is high, so that salts can be carried upwards in solution by capillary movement and deposited as the water evaporates, sands in the unsaturated zone may similarly contain the more soluble sulphates. Because of their aeolian method of deposition, dune sands are generally free of soluble salts but their fine and uniform grading makes them unsuitable for use in concrete other than as a minor replacement in the fine aggregate fraction. Their combination with contaminated sands therefore does little to reduce the total sulphate content of a mix.

The seriousness of the damage that has resulted from the use of sulphate contaminated aggregates has prompted a number of recommendations of a safe limit. The Cement and Concrete Association[2] recommends that the total amount contributed to the concrete from the cement, the aggregate and the mixing water should be not more than twice the amount present in the cement, which is generally about 2.0–2.5%. A maximum of 4.0% by weight of cement is specified in BS 5328.[4] Investigations into the causes of cracking in Middle East concretes and surveys of available materials led Fookes and Collis[5] to suggest a maximum limit of 0.4% for both fine and coarse aggregate with an overriding requirement that the concrete should not contain more than 4.0% by weight of cement. Samarai[6] found that mortar bars made with ordinary Portland cement in the proportion of 1:3, cement:sand, with added gypsum, did not expand more than 0.1%, the value he adopted as a tolerable level, at ages up to 36 months, when the total sulphate content did not exceed 5%. From these results he proposed that sulphates in aggregates should be limited to 0.6% when used with ordinary Portland cement. Test results on bars made with sulphate-resisting Portland cement indicated that 0.7% was a safe limit.

The source of Samarai's added gypsum consisted of anhydrous and partially hydrated forms of calcium sulphate, which may be presumed to have been a powder, and which he

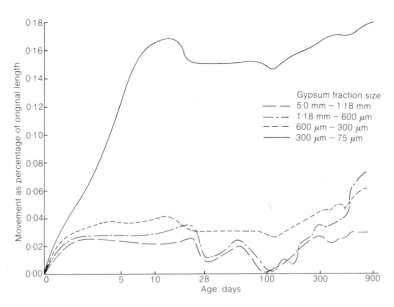

Fig. 1. Effect of gypsum fraction size on the expansion of mortar bars made with ordinary Portland cement (tricalcium aluminate content 12%) containing 6% of SO_3 and cured at 100% relative humidity

rehydrated in the mixing water and added to his mixes as a slurry. In practice, gypsum is likely to be present in contaminated aggregates throughout a range of gradings. The effect of particle size on expansion and compressive strength has been investigated in recent work[7] in which graded gypsum was added as a sand replacement in 1:3, cement:sand bars in amounts to give a range of sulphate contents as a percentage by weight of cement. At any one sulphate level the bars made with the finest gypsum fraction expanded most. Fig. 1 shows the expansion of bars containing 6% of sulphate. The compressive strengths of 75 mm cubes made with the same mixes as the bars were also measured. The strengths of those containing 4% of sulphate are shown in Fig. 2. It will be seen that at this acceptable level the influence of particle size is less marked. In their report the authors stressed that most published research on the effects of sulphates in aggregates has been related to the expansion of mortar specimens, and that this does not take into account the restraints imposed in concrete, particularly in reinforced concrete.

Recognition of a safe level for sulphate in concrete does not solve the problem of achieving it when aggregates of low sulphate content are not available. Washing, while effective in removing more soluble salts, is of little value in reducing the content of calcium sulphate, which is soluble in water only to the extent of 0.2%. Since solubility is a function of time of contact with the solvent as well as particle size, it is unlikely that aggregate washing plants could achieve the theoretical solubility of gypsum. Washing would remove any gypsum present on dusty surfaces of crushed rock aggregates, but would be without effect on sulphates within the particles. Similarly, although fine gypsum could be removed from sands, there would be little solution of coarser grains. Moreover, in arid regions fresh water is often scarce and costly, especially if it has to be tankered to site, so washing water is generally recycled and quickly becomes saturated with sulphates. This can happen within three to six cycles, depending on the original sulphate contents of both the water and the aggregates. In coastal areas the use of sea water without recycling might seem to offer an attractive alternative to washing with fresh water. However, its use not only would be ineffective in removing sulphates but also would have the very serious consequence of introducing chlorides into the concrete, as discussed later. Sea water is an ionically balanced solution. Its content of total dissolved solids varies with geographical location but the proportion of the ions present is constant. It is thus saturated with sulphate in equilibrium with its other constituents, and without a shift in the equilibrium is unable to dissolve more. Its use as wash water may therefore increase the sulphate content of the aggregates. Where only surface contamination results, as in the washing of materials of low porosity, the increase would be small, but it could be appreciable if the sea water was absorbed into porous aggregates. Total porosity values for limestones are given as ranging from 14 to 28%.[8] It can be calculated that the use of a highly saline sea water, containing perhaps as much as 3 g per litre SO_3, could increase the sulphate content of a porous limestone aggregate by up to 0.1%.

‹The use of sulphate-resisting Portland cement or of ordinary Portland cement blended with pozzolanic materials can reduce the risk of damage being caused by sulphate-contaminated aggregates." If chlorides are also present the use of sulphate-resisting cement may increase the risk of corrosion of reinforcing steel. This is discussed in the next section.

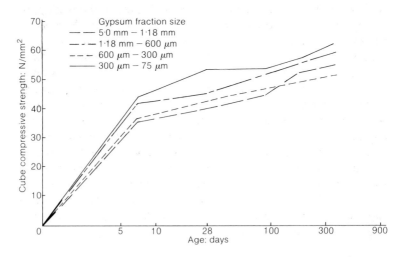

Fig. 2. Effect of gypsum fraction size on the strength of mortar cubes made with ordinary Portland cement (tricalcium aluminate content 12%) containing 4% of SO_3 and cured at 100% relative humidity

Concreting aggregates

Where aggregage deposits are known to contain gypsum, measures to control the content, by selection and frequent testing, should be instituted, both during production and on the construction site. Much can be done by selective digging, avoiding as far as possible obvious gypsum in surface deposits, in strata in a quarry face, and any pipes or chimneys in compact sand deposits. The avoidance of sulphates in loose and uniformly contaminated deposits, as, for example, beach sands, is more difficult. For quality control purposes it is unfortunate that the quantitative determination of sulphate is a fairly lengthy process and one which requires laboratory facilities and a certain degree of analytical skill. The introduction of instrumental techniques, such as nephelometry or turbidimetry, shortens the analytical procedures but not the time required for preparation of the sample and of the solution in which the final measurement is made. A staining test in which sulphates are identified by the co-precipitation of barium sulphate and potassium permanganate has been described.[9] This is a qualitative test, intended only to indicate the presence of sulphates, and its value is said to be in the positive identification it gives when sulphate minerals are present as fine grains and therefore are not readily distinguishable by visual means. Its usefulness in the testing of aggregates is more applicable in the field than in the laboratory, but even in the field its limitations should be recognised, particularly when it is applied to rock sources of aggregates. While a positive reaction for the presence of sulphate would indicate the need for caution, a negative result would not necessarily mean that the rocks were entirely free of sulphates. Where laboratory facilities are available, it is always preferable for the sulphate content to be determined quantitatively. It is sometimes possible to have the tests made by local testing organisations, but it should be stipulated that the sulphate content should be determined in a dilute hydrochloric acid extract of a ground sample. At present BS 812[10] does not include a method for the determination of the sulphate content of aggregates. Suitable methods are detailed in Chapter 7.

Chlorides in aggregates

Chlorides do not react expansively with Portland cements as do sulphates. Their effect when present in concrete is to increase the risk of corrosion of embedded metals of which the greatest volume used is steel reinforcement.[*] They can be tolerated in plain concrete, although when present in large amounts some surface dampness may result, but widespread and serious damage has been caused by the use of chloride-contaminated aggregates in reinforced concrete. Particularly at risk are thin stressing wires. The corrosion products occupy more than twice the volume of the steel and their formation can be accompanied by pressures as great as 32 N/mm$^{2\,11}$, resulting in cracking of the concrete, frequently followed by spalling of the cover. In severe cases of corrosion there may be a reduction in section of the reinforcing bars, leading to a loss of tensile strength of the concrete.

During the hydration of Portland cements, part of any chloride present reacts with the aluminates and ferrites to give the solid compounds tricalcium chloroaluminate and tetracalcium chloroferrite. It is stated that perhaps 90% of the chloride is so combined in ordinary Portland cement.[12] The chloride which remains in the pore water is responsible for corrosion of the steel.[*] It follows that in cements of low aluminate content less chloride can be combined as chloroaluminate and a greater proportion will remain in the free state. Corrosion can therefore be initiated at lower chloride levels in sulphate-resisting Portland cement concrete than in ordinary Portland cement concrete.[*] The aluminates cannot react with unlimited amounts of chloride, since the sulphate in the cement acts preferentially, and, the higher the sulphate content, the greater the amount of chloride that remains in the pore water.[a] A potentially dangerous situation thus exists when aggregates are contaminated with both sulphates and chlorides.[*] A further factor is involved when water is able to pass through the concrete. The

Table 5. Chloride content of unwashed Middle East aggregates

Aggregate type	Chloride (as percentage Cl)	
	Range	Mean
Carbonate rock coarse	0.02 – 0.17	0.12
Pit sand	0.01 – 0.98	0.32
Beach sand above water table	0.01 – 0.45	0.06

solid chloroaluminates and chloroferrites are in equilibrium with the free chloride in the pore water. If any of this is removed by leaching, chloride from the solid phases will pass into solution to restore the equilibrium and hence corrosive conditions will be maintained in the concrete.[13]

In the UK, although some land-won sands may contain small amounts of chlorides, the aggregates mainly likely to be contaminated are sea-dredged materials. Since mean sea water contains 1.8% of chloride (as Cl) the possibility exists for significant amounts to be left on the aggregates unless adequately washed. This was a matter of some concern when marine aggregates first became widely used, and in 1968 the Greater London Council issued a specification governing the use of marine aggregates in their works.[14] Further experience in the use of washed sea-dredged materials has shown that only exceptionally do they have chloride contents exceeding the limits given in standard specifications.

In arid countries the available aggregates, both sands and crushed carbonate rocks, may be heavily contaminated with chlorides. Sodium chloride is most prevalent but minor amounts of magnesium and potassium chlorides may also be present. Table 5 summarises the results of a number of testing programmes in several areas of the Middle East.

The limits for the chloride content of aggregates which have been fixed or proposed are shown in Table 6. It is unfortunate that two modes of expression have come to be used – chloride as chloride ion (Cl) and chloride as sodium chloride (NaCl). There is some merit in using Cl in line with the established practice of expressing sulphate as SO_3. The conversion factors from one to the other mode are:

Cl × 1.65   = NaCl
NaCl × 0.61   = Cl

For ease of comparison the conversions have been made in the limits given in the table, irrespective of how they were originally expressed.

Chlorides are appreciably soluble in water, to the extent of 35% in the case of sodium chloride, and can be readily removed from sands by washing. Removal is more difficult in the case of crushed rock aggregates where the salts are present throughout the particles. It has been found in laboratory tests that even after immersion for two hours in distilled water only 60% of the chloride was extracted from 5 mm limestone aggregate and as little as 30% from 25 mm stone. The efficiency of removal can be expected to be even less with the normal residence time in an aggregate washing plant. Nevertheless, washing of the sand size fraction can contribute significantly to reducing the chloride

Table 6. Limits for chlorides in concrete

Type of concrete*	Chloride (as Cl)	Chloride (as NaCl)	Percentage by weight of
BS 8110 (1985)[15]			
Prestressed	0.10	0.16	Cement†
Heat cured, containing embedded metal			
Plain, made with SRPC or SSC	0.20	0.33	Cement†
Containing metal and made with OPC, PBFC or combinations with PFA or GGBFS	0.40	0.66	Cement†
C & CA (1970)[2]			
Reinforced	0.06	0.10	Aggregate
Plain	0.40	0.66	Cement
GLC (1968)[14]			
Any concrete	0.06	0.10	Fine aggregate
	0.02	0.03	Coarse aggregate
Fookes and Collis (1976)[5]			
Reinforced	0.06	0.10	Fine aggregate
	0.03	0.05	Coarse aggregate
	0.30	0.50	Cement

* OPC = ordinary Portland cement (BS 12), SRPC = sulphate-resisting Portland cement (BS 4027), SSC = supersulphated cement (BS 4248), PBFC = Portland–blastfurnace cement (BS 146 and BS 4246), PFA = pulverised fuel ash (BS 3892), GGBFS = ground granulated blastfurnace slag.
† Inclusive of PFA or GGBFS.

concrete of the mix. Sea water must not be used for washing. Its high chloride content can only increase contamination.

The quantitative determination of chloride, in the fine aggregate at least is one that can be carried out on site (unlike the test for sulphate content) using simple test kits, of which a number are available. Figg and Lees[16] have described the use of a test strip. Other kits contain liquid or powder reagents and the measurements are made using simple methods and robust apparatus. However, all the kits are designed for the measurements to be made in aqueous solution, and the difficulty of removing all the chloride from coarse aggregate by the use of water is as pertinent to testing as it is to washing. In the tests mentioned above, complete extraction was not achieved until the samples had been ground to finer than 150 μm. It is possible that by careful control of conditions – particle size, ratio of sample to water, time of contact, temperature – a correlation could be established between total and extractable chloride which could serve as a site screening test. Other factors, such as variations in the porosity of the stone, would have considerable influence, and any site tests should be paralleled by frequent laboratory testing. A test method is given in BS 812[10] but, like that of Figg and Lees, it is designed particularly for those materials such as marine-dredged flint aggregates where the chloride is a surface contamination only. Laboratory tests on limestone coarse aggregates should therefore be made on finely ground samples.

Aggregate reactions with alkali metals

The alkali metals, sodium and potassium, are conventionally expressed in the results of chemical analyses of solids in terms of their oxides, sodium oxide (Na_2O) and potassium oxide (K_2O), although not present as such. In Portland cements they occur either as sulphates, or combined in the silicates and other minerals of the clinker. During hydration of the cement the alkali metal sulphates react with the aluminates and calcium hydroxide to give insoluble compounds and soluble sodium and potassium hydroxides which remain in the pore water, and it is these which enter into chemical reaction with aggregate constituents.

Because of the different atomic weights of the alkali metals it is difficult to equate the combined effect of their hydroxyl ion concentrations when they are presented in an analysis in terms of oxides, so a single convenient value has been adopted. This is total alkali content as equivalent Na_2O, and is the sum of the Na_2O plus the K_2O calculated as Na_2O. The calculation is made using the molecular weights of Na_2O (62.0) and of K_2O (94.2) and is:

Total alkalis as equivalent $Na_2O\% = Na_2O\% + 0.658 K_2O\%$

Alkali–silica reaction. The expansive reaction which occurs between certain types of siliceous aggregate and the alkali metal hydroxides, while of relatively recent occurrence in the UK, is known to have caused damage in the USA to concrete placed as early as 1914. By 1938 a number of bridges and roads in California were found to be affected by serious cracking, in some cases within only five years of construction, and in 1940 Stanton[17] published the results of investigations into the causes.

It was found that the fine aggregates which had been used in all the affected concrete were river sands of the same general type, containing shales, cherts and siliceous limestones. It was also found that the reactive sands only caused expansion when used with cements of alkali content greater than 0.6% equivalent Na_2O. Following the publication of Stanton's paper, investigations were made throughout the USA of concrete structures exhibiting the same type of distress and of the aggregates used in their construction. The minerals responsible for expansion were identified as opal, chalcedony, tridymite and crystobalite. Volcanic rocks of acid to intermediate composition including rhyolites, dacites and tuffs were also found to cause expansion. The reactive minerals contain silica in non-crystalline or crypto-crystalline forms and differ in reactivity according to the degree of disorder in their crystal structure. Opal, which is amorphous or highly disordered, is the most reactive. Chalcedony, tridymite, crystobalite, the glasses, and the crypto-crystalline and micro-crystalline forms of silica are less so. Strained quartz, which can occur in quartzite rocks and the sands derived from them, has also been found to be reactive, but quartz having a well-ordered crystal structure is unreactive.

The association of chalcedonic minerals with alkali reactivity and the fact that chalcedony occurs in the flint gravel aggregates widely used in the UK led Building Research Station (now Building Research Establishment) to conduct a programme of expansion bar and other tests on a number of UK aggregates, including crushed rocks as well as flint sands and gravels. The results were published in a series of National Building Studies papers.[18] It was concluded that although there were indications of reactivity when flint was used in certain combinations with other materials, none of the aggregates tested were expansively reactive when used as the whole aggregate.

The first recorded case of alkali–silica reaction in the UK was recognised in 1971 in a dam on Jersey and was attributed to the presence of opal and chalcedony in the diorite crushed rock and beach sand aggregates used.[19] The reaction was later found to have occurred in concrete in south-west England, South Wales and the Trent Valley of the Midlands of England. It was at first suspected that marine aggregates might have been responsible, since reactive flint sands and gravels dredged from the North Sea were known to have caused considerable damage in Denmark and north Germany. It was found, however, that only about half of the affected concrete had been made using sea-dredged material.[20]

The chemical processes involved in the alkali–silica reaction are fairly straightforward. The alkali metal hydroxides present in the pore water of the cement paste attack the reactive forms of silica to give a gel of alkali silicates of variable composition. Although calcium hydroxide does not take part in the attack, it can enter the alkali silicate gel from the pore water, and form alkali–calcium silicates. The reaction products absorb water, growing in volume and causing a swelling pressure to develop within the concrete. As more water is absorbed the gel becomes more fluid and able to flow into cracks and voids in the concrete and it may reach the surface and appear as wet spots or as exudations.

The factors which determine whether gel formation and swelling will result in sufficient expansion to crack the concrete are complex. There must be sufficient alkali hydroxide present to react with the silica and enough water available to promote swelling of the gel. The swelling is also influenced by the composition of the gels, those consisting essentially of alkali silicates being more expansive than those containing a high proportion of calcium.[21] The formation of gel is dependent on the presence of reactive silica in the aggregate but this need be present in only a minor amount. This critical content of reactive material has been termed the 'pessimum' value of an aggregate or aggregate combination. With highly reactive forms of silica, for example opal, a few per cent is sufficient to cause maximum expansion, while with slowly reacting aggregates the pessimum proportion is higher, and, in some cases, maximum expansion may not occur unless the aggregate consists entirely of reactive material.

The size of the aggregate particles which contain reactive silica is a factor in the expansion, since there must be sufficient reactive material at any point within the concrete to produce more gel than can be accommodated in the neighbouring pore space. Very fine particles of known reactive materials may not cause expansion. Indeed, Stanton found that an opaline chert, which caused damage when used in normal aggregate gradings, produced little expansion when finely crushed and used as a 15–20% cement replacement. In the damaged concretes examined in England the reactive silica was found mainly in the 1–5 mm particle size.[20]

Because of the complexity of the factors which produce expansion it is difficult to be categorical regarding aggregate types and combinations which might be susceptible. Rock types unlikely to be reactive are given[22] as those belonging to the Granite, Gabbro, Basalt (except for andesite), Limestone (except for dolomites), Porphyry (except for dacites, rhyolites and felsites) and Hornfels group classifications of BS 812,[10] but concrete made with innocuous coarse aggregate may still expand if the sand used contains reactive material. In some of the affected concrete in the UK the coarse aggregate was limestone or granite and expansion was caused by the reactivity of the flint fine aggregate used. Early work in the USA suggested that in general significant expansion of concrete made with flint aggregate would only occur when flint was used in particular proportions with inert materials. Jones[23] considered that flint gravels and sands should strictly be regarded as being slightly reactive, although when used as the whole aggregate they are normally unlikely to cause expansion. Currently there are no British Standard test methods for determining the alkali reactivity of aggregates, ASTM methods being generally followed. These are discussed in Chapter 7.

• Although a different mechanism is involved in it, an expansive reaction can take place between alkalis and certain silicate rock aggregates. The reaction was first recognised in Nova Scotia, where the natural aggregates consist of sedimentary rocks including phyllites, which are intermediate between slates and schists in mineral composition and grain size and are placed in the Schist group classification of BS 812. The reaction differs from that of the alkali–silica reaction, and for that reason is sometimes referred to as alkali–silicate reaction. In investigations made in Canada the expansive reaction was attributed to exfoliation of clay minerals in the rocks, not to gel formation.[24]

Alkali–carbonate reaction. The alkali–carbonate reaction is of rare occurrence and has not been proved to have been a cause of damage to concrete outside North America. It was first investigated in Canada in 1955 and an early report was given by Swenson in 1957.[25] Widespread expansion and cracking of concrete had been found, mainly in structures such as pavements, kerbs, floors and foundation walls, all of which were exposed to damp conditions. The cracking followed the pattern characteristic of alkali–silica reaction, but certain features, including gel exudations, were absent. Although chemical tests indicated non-reactivity of both coarse and fine aggregates, concrete prisms made with a high alkali cement showed expansions of 0.25% within six months storage at 100% relative humidity. The coarse aggregate, which was a fine-grained argillaceous dolomitic limestone of Ordovician age, was identified as the reactive material.

Further investigations of carbonate rocks found that those which were expansive had similar compositional and textured features. In composition they were of intermediate calcite to dolomite ratio and all contained appreciable amounts of clay or clay-sized minerals, but not highly swelling clays. The characteristic textual pattern was of small rhombic crystals of dolomite, less than 50 µm across, set in a fine-grained matrix of calcite and clay. There was a tendency for the clay to be concentrated around the dolomite rhombs with some apparent inclusions within the crystals. Although characteristic of reactive rocks, the same features were found in some rocks which did not expand in the presence of alkalis.

• The mechanism of alkali–carbonate rock reaction is complex and disputed. It is agreed that the first stage is attack by the alkali metal hydroxides on the dolomite crystals, giving calcite, brucite (hydrated magnesium oxide) and alkali carbonates – a process which has been termed dedolomitisation because, in effect, it destroys the dolomite in the rock. There is no general agreement on whether the chemical reaction is directly responsible for expansion or whether it has an indirect effect in that it provides conditions favourable for other reactions to occur. Gillot and Swenson[26] advocate the indirect mechanism and argue that if the chemical reaction alone were responsible all dolomitic rocks would cause expansion when used as concrete aggregates. They consider that the prime cause of expansion is due to clay, which is present in reactive rocks. They propose the theory that the water which was held in the lattices of the clay minerals at the time when the sedimentary deposits were laid down was later expelled by consolidation and other rock-forming processes. After lithification of the deposits, water was unable to penetrate into the fine-grained rocks to rewet the clay minerals, unless channels for its entry were opened by chemical or physical action. It is suggested that in concrete, microcracks, resulting from the dedolomitisation of the aggregate, provide the necessary channels. On rehydration, the desiccated clay minerals increase in volume and the swelling pressures thus created cause the concrete to crack. If this explanation is accepted, and bearing in mind that in Canada the reactive rocks were only found at considerable depths in the quarries, alkali–carbonate reaction cannot seriously be considered to be a cause of damage to concrete made with dolomite or dolomitic limestone rocks of evaporite origin, such as occur in the Middle East.

Reaction rims often form around aggregate particles in concrete made with crushed carbonate rocks, but studies have shown that they do not give a positive indication of alkali-carbonate reaction having taken place. Aggregates developing rims were generally found to be similar in composition to the reactive rocks, but not all proved to be expansive. Conversely,

Iron minerals in aggregates
Minerals which contain iron in the ferrous state and which readily undergo oxidative weathering are undesirable in concrete aggregates. The minerals chiefly involved are the iron sulphides, pyrite, FeS_2, and pyrrhotite, or magnetic iron pyrite. The sulphides occur in basic igneous rocks such as gabbros and in contact metamorphic deposits, including metamorphic limestones. Iron pyrite is also found in flint gravels. In general, oxidation results in no greater damage than rust staining and popping of concrete surfaces, but in extreme cases disruption of the concrete may occur because of attack on the cement paste by sulphates.

The oxidation of pyrite takes place in the presence of moisture and atmospheric oxygen. Ferrous sulphate and sulphuric acid are first formed, but since ferrous sulphate in solution or under moist conditions is stable only in the presence of sulphuric acid, if the acid formed can be neutralised, as for example when pyrite is present in concrete, a second reaction follows. In this, the ferrous sulphate is oxidised to rust-brown ferric hydroxide, which is stable under atmospheric conditions and is the cause of staining.

The chemical reactions do not inevitably occur when pyrite is present in a concreting aggregate. Midgley[27] investigated the staining of concrete made with Thames river gravels containing pyrite, and recognised two forms, reactive and unreactive, distinguished by their behaviour in saturated lime water. A few minutes after immersion the reactive varieties produce a green-blue precipitate of ferrous hydroxide which rapidly changes to a brown precipitate of ferric hydroxide. He found no relation between the physical forms of pyrite and its reactivity, but chemical and spectrographic analyses showed that the unreactive forms contained more other metals (antimony, arsenic, copper, lead, manganese, tin, zinc) than did the reactive varieties. The analyses also showed that both forms contained less sulphur than the exact two atoms of the theoretical formula. This is not unusual since pyrite has a defective structure and rarely contains stoichiometric proportions of iron and sulphur, the latter usually being deficient. Midgley concluded that the extra metals in the unreactive pyrite in some way stabilised the structure so that the reaction with lime water did not occur.

A different mechanism has been proposed to explain the rapid oxidation of pyrite present in carbonaceous shales of middle-Cambrian, lower Ordovician origin in Norway.[28] This was attributed to the presence in the shales of a rare, monoclinic form of pyrrhotite which was thought to accelerate oxidation of the pyrite, possibly by an electro-chemical reaction. The shales are not satisfactory concrete aggregates because of their physical properties. The problems they present are sulphate attack to concrete placed in contact with them, but where they have been used as aggregate in concrete made with ordinary Portland cement severe disruption has occurred as a result of sulphate attack.

The presence of pyrite and pyrrhotite in aggregates is also reported to have caused sulphate attack to concrete in Sweden.[29] In one dam structure the presence of up to 10% pyrrhotite in the aggregate resulted in extensive swelling and cracking of the concrete which had been made with ordinary Portland cement. This is not surprising, since the complete oxidation of pyrrhotite produces roughly the the same amount of sulphate (as SO_3). It has been recommended in Sweden that about 1% should be the upper limit for easily oxidisable sulphide minerals, especially iron pyrite, in aggregates.[30]

The oxidation of ferrous iron in biotite mica in igneous rocks can result in staining. A spectacular example is shown in the pattern of weathering of Hong Kong granite.[31] It is also considered that the oxidation of ferrous iron present in ferromagnesian minerals contributes to the deterioration of some concretes made with dolerite aggregates. The readily oxidisable minerals occur as degradation products of olivine, and there is evidence that oxidation can be accompanied by expansion.[32]

Although the ferrous forms of iron may undergo oxidative weathering, in its ferric form iron is in its highest state of valency and is resistant to atmospheric oxidation. Nor does it react with any cement constituents to cause staining or expansion. It is therefore difficult to understand why limits should sometimes be placed on the amounts of ferric compounds permitted in job specifications for aggregates. This is particularly so in the case of magnetite, Fe_3O_4, which is found in igneous rocks as an accessory mineral and in other rock types. Magnetite is highly resistant to weathering processes, either chemical or erosive, and when found as individual grains, it represents the residual product of the alteration of the parent rock. There can be difficulties when ores, including magnetite ore, are used as heavy aggregates in concrete for nuclear shielding, but these arise not from any chemical reactions, but from the physical properties of the ores, which make it difficult to obtain suitable aggregate gradings to allow adequate workability in the mix without segregation.[33]

Organic matter
Certain types of organic matter, if present in significant amounts in aggregates, can retard the rate of setting and hardening of Portland cements. The active substances are present in the humic compounds derived from the decay of plant and animal tissues, and are most often associated with silt and clay-sized material in pit sands. A distinction must be made between the presence of recognisable particles of incompletely decomposed plant material, such as lignin, whose effect, if any, is to cause staining of concrete surfaces, and the true humic colloids containing the chemical compounds which retard the set of Portland cement.

Soil organic matter is chemically very complex, since it is subject to continual processes of microbial degradation and synthesis. During these processes readily assimilable and therefore relatively short-lived compounds are formed. These have definite physical and chemical properties and include sugars, proteins, fats, waxes and low molecular weight organic acids. The residual humic substances, which are intermediates in the conversion of plant residues into carbon dioxide and inorganic salts, have no specific chemical or physical properties, but consist of fairly stable, dark-coloured, amorphous material, which has been estimated to have a mean residence in the soil of 1000 years.[34]

Three main fractions of humic material are recognised – humic acid, fulvic acid and humin. The three fractions are thought to be structurally similar, differing only in respect of molecular weight, elemental analysis and content of functional groups.[35] The principal functional, or active, groups are carboxylic acids and phenolic hydroxides, but up to 25% of the carbon in humic matter is contained in simple sugars. Carboxylic acids are the active agents in a number of commercially prepared retarding admixtures for use in concrete, while it is the presence of sugars which confers retarding properties on lignosulphonate based water-reducing and retarding admixtures.

Because of the complex composition of humic matter, chemical tests to determine deleterious amounts in fine aggregates are non-specific. For many years BS 812[10] included a test based on the depth of colour developed when the sand was left in contact with a solution of sodium hydroxide, but it was considered that the test suffered the disadvantage that some compounds developing a colour could be without effect on the setting time of cement, while the presence of deleterious compounds might not be detected. The colour test was replaced by the measurement of pH values of cement–sand pastes, but this test was also withdrawn when experience found that not all Portland cements gave a positive response to the presence of organic matter. Currently there is no British Standard chemical test for organic matter. Strength tests on concrete are recommended in BS 882[36] if there is reason to suspect that impurities in aggregates may result in acceleration or retardation of the set of the cement or in air entrainment. A colorimetric test is retained in ASTM C40-73[37] but it is stated in the method that the test is only an approximate determination of injurious organic compounds, and that its principal value is to furnish a warning that further tests on suspect sands may be necessary. A further test which is recommended is given in ASTM C87-69 (1975).[38] This is the comparison of the 7 day compressive strengths of mortar cubes made with the sand in its natural state and after treatment with sodium hydroxide solution until its organic content has been reduced to the acceptable level of test C40-73. However, fine aggregates which fail the colorimetric test may be used provided that the colour is due principally to small amounts of coal or lignite. Where appprearance is important the content of these is limited to 0.5%. For other concretes the maximum allowed is 1.0%.[39]

There is some evidence that the fulvic fraction of soil organic matter most actively retards the hydration of Portland cements. It has been found that fulvic acids separated from a variety of soil types contained twice as much of the organic compounds identified as retarders as did the humic fractions of the same soils.[40] The activity of fulvic acid has been recognised in Europe, and a colorimetric test which measures only the fulvic fraction has been adopted by one manufacturer as a routine test for sands to be used in precast concrete products.[41]

The staining caused by lignin in pit sands has been mentioned. Staining due to reaction between the alkalis present in Portland cements and organic matter in limestones may also occur. The disfigurement is more apparent in masonry construction than in concrete, and takes the form of irregular dark staining at the edges of the blocks at the mortar joints. It has also been known to occur where fine-grained limestones have been used as decorative cladding panels, and in artificial stone units made with limestone aggregate. In the latter case the staining is more uniform.

ARTIFICIAL AGGREGATES

The term artificial refers here to materials which are natural in origin but have been processed in some way, generally by heat treatment, before use as aggregates. Included are by-products, such as blastfurnace slag, pulverised fuel ash and clinker and natural materials which are used as lightweight aggregates.

Dense aggregates

Air-cooled blastfurnace slag is the main dense artificial aggregate and it has been used extensively in concrete. Other materials, less used, include crushed fired clay bricks and burnt oil or colliery shales, the latter particularly in the production of concrete bricks and blocks.

For use as dense aggregate, blastfurnace slag is allowed to cool slowly in air, when it forms a crystalline solid. The general chemical composition of slags was given in the previous chapter. The mineral phases which form on cooling have been described by Nurse and Midgley.[42] Requirements for the chemical properties of blastfurnace slag aggregate are given in BS 1047.[43] The requirements are that the slag should be free from 'falling' or 'dicalcium silicate' unsoundness and from 'iron' unsoundness and should not contain more than specified amounts of sulphur compounds. 'Falling' is due to the presence of a metastable form of dicalcium silicate, larnite, which, on reverting to its stable form, expands and causes the slag to disintegrate. To determine whether the composition of a slag is such that larnite may be a constituent mineral, an equation derived from phase equilibria is applied to the results of chemical analysis. If the equation indicates that larnite is likely to be present, confirmation or otherwise is given by microscopic examination of a polished and etched specimen.

Iron unsoundness is a rare disorder, and, as far as is known, does not occur in the UK in production slags, although it has been induced experimentally by the addition of millscale to molten slag. Iron unsound slags readily break down in water, and this behaviour is the basis for the test given in BS 1047. Pieces of slag are immersed in water for 14 days, at the end of which time any particles showing 'cracking', 'disintegration', 'shaling' or 'flaking' are deemed to be iron unsound. Unfortunately, many interpretations have been put on these terms. For example, it has been argued that dust dislodged from the pores of the particles during the test indicates flaking. It has also been thought that the leaching of sulphur as yellow-green polysulphides, which is quite common, indicates the disintegration of a constituent mineral. Unsound slags break down in a very characteristic way, opening in layers from the surface inwards, in what may be described as an onion peeling effect.

Blastfurnace slags all contain sulphides, chiefly as calcium sulphide, although some may be present as iron or manganese sulphides. A small proportion may oxidise to sulphate during cooling. The limits for their contents given in BS 1047 are that the total sulphur content should not exceed 2% (as S) and that the sulphate content should not exceed 0.7% (as SO_3).

It was at one time thought that the sulphides in blastfurnace slags would promote corrosion of steel reinforcement, and that slag aggregates should not be used in reinforced concrete. This view probably arose because other sulphur-bearing materials, such as coke breeze, had been found to cause corrosion. Although blastfurnace slag aggregate had a good service record in reinforced concrete, reservations about its use persisted, and these led Building Research Station to make a study of the corrosion of steel in concrete made with slag aggregates. Tests were carried out on the weight loss of steel plates immersed in aqueous extracts of powdered slags and of steel rods embedded in concrete bars made with slag coarse aggregate, exposed outdoors for two years.

The conclusions were that, whereas theoretically the sulphides should influence corrosion, the high alkalinity of the concrete, due to both the cement and the slag, provided a protective environment in which corrosion of the steel was restricted to negligible amounts.[44]

Lightweight aggregates
Blastfurnace slag is also processed to produce lightweight aggregates in the form of crushed foamed (expanded) slag or in the form of pelletised expanded slag. In the foaming process, the molten slag is treated with a limited amount of water, and the gases, mainly steam, which escape from the hot mass, leave a vesicular structure in the cooled slag. In the pelletising process, the molten slag is expanded by water sprays, then passed over a cooled, spinning drum fitted with vanes that break up the still hot and plastic material and throw it into the air for sufficient time for pellets to form through surface tension effects. In either form, expanded slag for use as aggregate is mainly crystalline. The glassy pellets produced for cementitious use are formed under different operating conditions. The properties of expanded blastfurnace slag aggregates are governed by BS 877.[45] The chemical requirements are that the sulphate content should not exceed 0.1% (as SO_3) and that the aggregates should be free of any other contaminants which might affect the durability of concrete made with them. A limit for total sulphur content, as stipulated for dense slag aggregate, is not necessary for expanded slags, since the sulphides are largely lost in the foaming process.

A number of lightweight aggregates are included in BS 3681.[46] These are exfoliated vermiculite, which is a clay mineral heated to release the water held in the lattice, so expanding the clay layers; expanded perlite, which is a glassy mineral of volcanic origin, expanded by rapid heating to incipient fusion point; expanded clay and shale and sintered pulverised fuel ash. Natural pumice is also covered by the specification. This does not require heat treatment, and strictly does not come within the definition of artificial aggregates adopted here. Pumice is of volcanic origin and is a glassy honeycombed mass. It might be thought that pumice and perlite would enter into alkali–silica reactions, but this appears not to be the case. The sulphate contents of all types of aggregate included in the specification is limited to a maximum of 1.0% (as SO_3). The loss on ignition is limited to 4.0%, with the exception of exfoliated vermiculite, for which a value is not stipulated.

Clinkers and other furnace residues are by-products used as lightweight aggregates and they are governed by BS 1165.[47] Clinker results from the burning of lump coal and is defined as well-burnt furnace residue which has been fused into lumps. The residue from furnaces fired with pulverised fuel is called furnace bottom ash. Neither type is recommended for use in concrete containing embedded metal or in concrete required to have high durability. The chemical requirements are that the sulphate content should not exceed 1.0% (as SO_3), and that the loss on ignition should not exceed 10% in aggregate for use in concrete for general purposes, or 25% in aggregate for use in concrete in interior work where conditions are mainly dry. The specification draws attention to the need for discretion in selecting and using clinker aggregates.

4
Other concreting materials

Previous chapters dealt with the chemical composition of the two main constituents of concrete, the cement and the aggregate. Another essential component is water. Other materials may be used to modify the behaviour of the cement or to confer special properties on the concrete. These include admixtures and polymers. Cement replacement materials, such as pulverised fuel ash and ground granulated blastfurnace slag, are here not considered to be admixtures but to be cementitious components of the concrete.

WATER FOR CONCRETING

The quality of the water to be used for mixing or curing concrete is generally taken for granted, as indeed it can safely be where adequate supplies of water of potable composition are available. This is usually the case in temperature climates but in arid regions of the world waters may contain high levels of dissolved salts which can lead to deterioration of the concrete. Even in countries well served by a public distribution system of potable water, circumstances may arise where water for concreting must be drawn from untreated sources, and it is then necessary to ensure that it does not contain harmful impurities. Surface waters may contain organic substances such as humic matter which can have a retarding effect on the setting time of the cement, or the growth of certain forms of algae may result in the release of chemical products which have the property of entraining air, with a consequent reduction in the strength of the concrete. Ground waters may contain undesirably high levels of dissolved salts.

Fairly high levels of dissolved inorganic constituents are

generally tolerable in mixing water, and guidance on contents of sulphates, chlorides and alkali carbonates and bicarbonates is given in BS 3148:1980.[1] The most critical factor is the chloride content of water to be used for making reinforced concrete. A maximum content of 500 mg Cl per litre is given as a guide in BS 3148, but it is recommended that the amount present in the mixing water should be considered in relation to that present in the other materials and that the total chloride content of various types of concrete should not exceed the limiting values given in CP110:Part 1:1972, now superseded by BS 8110:Part 1:1985.[2] These limits are shown in Table 1 of reference 2. Sea water, containing 18 000 or more mg per litre of chloride, has been successfully used for mixing plain concrete made with ordinary Portland cement, but it should not be used in reinforced concrete, because of the risk of corrosion of the steel. It should not be used for mixing either plain or reinforced concrete if the aggregates contain alkali-reactive forms of silica, since chemical interactions between hydration products of the cement and the sodium and chloride ions present in the water can result in an increase in the alkali content of the mix.

Similarly, large amounts of the alkali metal, sodium and potassium, carbonates and bicarbonates may contribute to alkali–silica reactions, and the specification suggests safe limits for their content in the mixing water. They may also affect the setting time of the cement. The alkali metal carbonates and bicarbonates are uncommon constituents of UK waters, but are known to be present in significant quantities elsewhere. In the UK they may be found in waters derived from, or which have passed through, geological formations containing minerals which have a cation exchange, that is, a water softening capacity. The minerals which have this effect are generally glauconites, which occur in the Lower Greensands.

The pH value of mixing water is, perhaps surprisingly, not very important, at least within the range of pH 4.5–9.0. Any acidity, due for example to dissolved carbon dioxide, is quickly neutralised by the alkalinity of the cement. Values lower than 4.5 in natural waters may suggest the presence of humic acids which can retard the setting time of Portland cements, and values in excess of 9.0, the presence of alkali carbonates. Chemical analysis of the water will reveal the presence of the latter, but if humic substances are suspected, physical tests on cement or concrete specimens are necessary. Tests for the determination of the setting times of cement and for the compressive strength of concrete cubes are recommended in BS 3148 to assess the suitability of water for mixing concrete, and tolerable reductions in strength and delays in setting are indicated. These tests should be made whenever there is any doubt about the quality of the water. However, untreated waters, surface waters in particular, are subject to seasonal changes in composition. On long contracts, initial testing may not provide sufficient assurance regarding the continuing quality of the water, and further testing is advisable whenever changes are noted in its physical condition. These may be recognised as changes in colour or odour, or as the development of algal or other organic growths.

The quality of the water to be used for spray curing of concrete is as important, if not more important, than that of the mixing water. Dissolved salts can readily penetrate into the green concrete and accumulate within it as the water evaporates in the form of vapour. Reinforced concrete is specially vulnerable to the effects of chlorides, and even plain concrete may suffer surface spalling if salts, particularly sulphates, which have been introduced in the curing water, migrate towards the surface and crystallise below the skin. Sea water can be used for curing plain concrete if any efflorescence or surface dampness which may develop from the chlorides present can be tolerated but it should never be used for curing reinforced or prestressed concrete.

ADMIXTURES FOR CONCRETE

Admixtures for use in concrete are chemical compounds, mainly organic or partly organic in nature, which are used to modify the properties of plastic or hardened concrete. The specific uses of different types are to accelerate or retard the normal setting time of the cement, to increase the workability of the freshly mixed concrete, to allow a reduction in the water/cement ratio used in the mix thus either increasing the strength of the hardened concrete or allowing a reduction in cement content for a given strength requirement, to entrain air and to act as integral waterproofers. Although the effect on Portland cement of a number of the compounds currently in use had been known for many years, the commercial exploitation of the organic admixtures did not begin until the mid-1950s. They were at that time viewed with some suspicion on the grounds that they were no substitute for good concreting practice, and, indeed, that is still true. In addition, some of the early admixtures, notably the lignosulphonates, were unrefined products of variable composition and inconsistent in their effects. In some early work on admixtures, mainly lignosulphonates, it was found that for a given time of retardation of set the required dosages of the products tested varied between 80 and 500% of those recommended by the manufacturers.[3] Since most organic admixtures are used in very small amounts, generally less than 1% by weight of cement, quite small variations in dosage can have marked effects. Their action is also influenced by the composition of the cement with which they are used.

Admixtures for specific purposes are covered by BS 5075.[4] Part 1 of the specification defines five types – accelerating, retarding, normal water-reducing, accelerating water-reducing and retarding water-reducing. Part 2 covers air-entraining admixtures and Part 3 covers superplasticising and retarding superplasticising materials. The three parts of the specification indicate the types of chemical substance which may be used in each group of admixtures. The main active constituents of a number of proprietary materials are included in a data sheet[5] and the chemical structures of the compounds most commonly used and their reactions with cement have been described by Rixom.[6]

The functions of accelerating admixtures are to increase the rate of setting of the cement and to promote early development of strength in the concrete. They are used in cold weather working when the rapidity of set and the high early strength provide the fresh concrete with increased resistance to frost action, and are also used to allow the early removal of formwork and the early demoulding of precast concrete products. A number of compounds, mainly inorganic salts, have been found to act as set accelerators and are included in proprietrary products. Calcium chloride has been the most widely used. Accelerating admixtures increase the rate of hydration of the cement, with the result that the heat evolved quickly builds up within the concrete. It has been stated[7] that the addition of 2% of calcium chloride by weight of cement has the same effect on accelerating hydration as a rise in atmospheric temperature of 11°C, irrespective of the ambient temperature at the time of concreting.

Since calcium chloride has been the most widely used set accelerator, its reaction with cement has received most attention but the mechanism by which it modifies the normal hydration process is not known with certainty. Ramachandran[8] considered a number of proposed theories but concluded that none explained all the effects. In the hardened cement paste, most of the chloride is combined as calcium chloroaluminate and chloroferrite. This is of greater significance to their role in corrosion than to hydration, in which, in any case, they might be expected to have a retarding effect in the same way as the formation of calcium sulphoaluminate delays the set of Portland cements retarded with gypsum.

A great deal of damage has been caused by the corrosion of steel when calcium chloride has been used in reinforced concrete, even though not in excessive amounts. The quality of the concrete and its conditions of exposure are important factors in determining whether corrosion will occur. Calcium chloride is also considered to increase the expansion of concrete caused by alkali – silica reactions[9] and to reduce its resistance to attack by sulphates.[10] The use of chloride-containing admixtures is covered by BS 8110.[2] The maximum chloride content of the admixture is limited to 2% by weight. The maximum amount which its use may introduce into the concrete is 0.03% chloride by weight of cement for concrete containing embedded metal and made with any type of cement and for concrete without metal made with sulphate-resisting Portland cement or with supersulphated cement. These stipulations, although applying to all types of admixture, restrict the use of calcium chloride to plain concrete made with ordinary Portland cement, either alone or in combination with ground granulated blastfurnace slag or pulverised fuel ash. The permitted dosage is dependent on the amount of mixing water used. For a water/cement ratio of 0.5, the maximum allowed would be 1.5% of anhydrous calcium chloride by weight of cement.

The other chemical substances which are known to act as set accelerators have found less general use than calcium chloride, mainly because of cost, but also in some cases because their erratic behaviour, except under carefully controlled laboratory conditions, makes them unsuitable for site use. Salts of formic acid, principally calcium formate, are increasingly used since they do not present a risk of corrosion to reinforcing steel. Solutions, or mixed solutions, of sodium aluminate, carbonate and silicate are used as accelerators and are of particular value when a very rapid set is required, as, for example, when concreting or grouting under water. The organic compound, triethanolamine, has some use as an accelerating admixture, but above a certain concentration in the concrete it acts as a retarder.

The retarding and water-reducing admixtures are mainly organic compounds. There is some overlapping in their effects, so the same compounds can be found in both groups of admixture. Their main active ingredients belong to one of three types – lignosulphonates, hydroxycarboxylic acids and carbohydrates. The chemical names may seem formidable, but the compounds are either naturally occurring, familiar substances or are derived from natural materials.

The lignosulphonates are produced from the spent liquors of chemical wood pulping processes in the form of calcium, sodium or ammonium salts of lignosulphonic acid. The active retarding agents are sugars, which vary in type and content, according to the source of the lignosulphonate liquor and the refining process. Used alone, the lignosulphonates act as retarders. For use as normal water-reducers only, the retardation is counteracted by the addition of set accelerating agents, such as triethanolamine. Lignosulphonates can also entrain air, an effect which is overcome by the addition of air-detraining, or antifoaming, materials such as dibutyl phosphate, various alcohols or silicones. The hydroxycarboxylic acids are naturally occurring organic acids belonging to a group which includes citric and tartaric acids, although other members of the group are more commonly used in admixtures. The admixtures are usually salts of the acids, generally the sodium salt. At a low dosage, these materials act as normal water-reducing admixtures. Used at a high dosage they are also retarding. The carbohydrates, also called polyhydroxy compounds, are derived from materials such as starches which are chemically treated to give less complex compounds. They act either as retarders when used alone,

Other concreting materials

or, with the addition of a set accelerator, as normal water-reducing admixtures.

The materials used in superplasticising admixtures are purified lignosulphonates and salts of the synthetic compounds sulphonated melamine formaldehyde and sulphonated naphthalene formaldehyde. Their special properties are to give very high workability or to allow a very large reduction in the water content of the concrete for a given workability, and to achieve these, the admixtures are used at much higher dosages than are the retarding and water-reducing types. The sulphonated formaldehyde compounds can be used alone at the high dosages without entraining air, the lignosulphonate admixtures generally contain other constituents to reduce air-entrainment.

The action of all types of retarding and water-reducing admixture on cement is complex, but is considered to be one of absorption on to the cement and its early hydration products. Under normal conditions, the hydration products initially form a suspension which changes to a stiff network as hydration proceeds. The retarding compounds lengthen the time of the suspension stage, thus delaying the onset of stiffening. This has also been described as deflocculation of the cement particles caused by repulsive electrostatic charges which the absorbed admixture imparts to the surface of the hydrating cement.[11] The initial plastic viscosity of the cement paste is reduced, allowing a reduction in the amount of mixing water required to obtain a given workability.

The materials used when air entrainment is required in the concrete are mainly wood resin soaps or synthetic surfactants. Their function is to lower the surface tension of the mixing water so that foaming will occur, and they are not involved in any chemical reactions with the cement.

Other compounds which may be added to concrete are those termed integral waterproofers. They fall into two main groups – those which react with the cement to form insoluble compounds in situ and those which are used in an insoluble form and which serve as hydrophobic fillers in the pores of the concrete. The reacting waterproofers are mainly soluble soaps of stearic or other fatty acids which combine with the calcium hydroxide of the cement to form insoluble calcium soaps. Soluble sodium and potassium soaps are prone to froth, and have come to be replaced by butyl stearate which reacts in the same way in the cement but without frothing. The hydrophobic fillers include calcium and aluminium soaps, plant and mineral waxes and resins. Integral waterproofers are not much used in concrete, but may be of benefit in structures exposed to one-sided, low water pressures, as in retaining walls, basements and tunnels. The term waterproofer should not be taken literally, since their use does not stop the passage of water through porous concrete nor do they prevent the movement of water vapour.

POLYMERS IN CONCRETE

From their structural formulae, many of the organic materials used as plasticising admixtures in concrete can be described as polymers. Their function, however, is merely to reduce the permeability of the concrete by allowing a lower water/cement ratio to be used, and although this improves the chemical resistance of the concrete, the admixtures do not provide direct protection. What are termed polymer cement concretes contain fairly large amounts, about 10% by weight of cement, of various organic monomers and catalysts. Polymerisation takes place within the concrete and the hardened polymer protects the hydrated cement against chemical attack. The addition of the monomers also increases the plasticity of the mix and allows a reduction in the water/cement ratio. Hardened concrete may also be impregnated with polymers to reduce its pore volume. In this technique the monomers and catalysts are injected into the dried concrete under pressure. It is therefore more suitable for precast than for site concrete, although it may have application in repair works. The addition of polymers by either method considerably increases material costs and they are likely to find only specialised use.

The materials called polymer concretes do not contain hydraulic cement. The binders consist entirely of polymers, such as polyester or epoxy resins. They are expensive and their use is mainly confined to providing chemically resistant floors or linings.

5
Chemical attack on concrete from external sources

SULPHATES

The damage that can be caused to Portland cement concretes by sulphates present in aggregates has been described in Chapter 3. The mechanism of the chemical reaction was confined there to that of calcium sulphate (gypsum) which is the most usual sulphate contaminant of aggregates. It is also the most common type of sulphate occurring naturally in soils, and it occurs in the UK in the London Clay, Lower Lias, Oxford Clay, Kimmeridge Clay, Weald Clay, Gault Clay and Keuper Marl.[1] Sodium sulphate (Glauber's salt) and magnesium sulphate (Epsom salt) may also be present. These occur in the UK in limited formations only, but extensively in other parts of the world, as in the so-called alkali soils of North America, and in the saline soils of the Middle East and other arid areas. These salts have a far more damaging effect on Portland cement concrete than does gypsum, since not only are they more soluble, allowing greater concentrations of sulphate in the ground water, but also they enter into more complex chemical reactions with the cement minerals, reactions which can result in complete deterioration of the cement paste. Ammonium sulphate, which is particularly aggressive, does not occur naturally. Its action on concrete is described in the section below on Industrial chemicals. The natural occurrence of iron sulphates, due to the oxidation of pyrite, is discussed below under Acids of natural occurrence.

The action of calcium sulphate on Portland cements is relatively simple. It reacts with the tricalcium aluminate and, to a lesser extent, the tetracalcium aluminoferrite phases, to give mainly tricalcium sulphoaluminate (ettringite) and calcium hydroxide.

The action of sodium sulphate is two-fold. It reacts first with the calcium hydroxide liberated during hydration of the cement, forming calcium sulphate and sodium hydroxide. The calcium sulphate acts on the tricalcium aluminate to give ettringite, in the same manner and with the same consequences as outlined above. However, in flowing water, where there is continual replenishment of sodium sulphate to the surface of the concrete and continual removal of the sodium hydroxide produced, the first reaction is self-perpetuating and can proceed until all the calcium hydroxide has been converted to gypsum. In static waters, the reaction is limited, depending on the concentration of sodium sulphate in solution.[2]

Magnesium sulphate has a much more damaging effect than either calcium or sodium sulphate, since the chemical reactions into which it enters destroy the calcium silicates which are the major phases present in the cement paste. Its initial reactions are with the tricalcium aluminate to give ettringite and magnesium hydroxide and with the calcium hydroxide to give calcium sulphate and magnesium hydroxide. Magnesium hydroxide is less soluble than calcium sulphate, hence the pH value of its saturated solution is lower than that of the saturated solution of calcium hydroxide which is the aqueous phase in the cement paste. The magnesium hydroxide causes the pH value of the pore water to drop below the level required to stabilise the calcium silicates, which then release calcium hydroxide into solution to re-establish their equilibrium pH value. This calcium hydroxide in turn reacts with magnesium sulphate, producing further magnesium hydroxide which again lowers the pH, and the silicates again release calcium hydroxide to raise it.

If sufficient magnesium sulphate is present, the reactions can continue to the complete breakdown of the calcium silicates to silica gel only. The ettringite formed in the reaction with the aluminates also becomes unstable when the pH drops, and in the presence of magnesium sulphate breaks down to calcium sulphate, aluminium hydroxide and magnesium hydroxide.

The calcium sulphate formed in the various reactions remains in solution until saturation point when it crystallises out as gypsum.

In temperate climates sulphates generally become leached from the top metre or so of soil, and concrete placed at a shallow depth and above the highest level of the water table is at little risk from sulphate attack. However, water levels vary seasonably, and if a normally high water table has been lowered during periods of prolonged dry weather, after rainfall it may greater depths.

In hot, arid countries sulphates may be carried upwards by the capillary movement of ground water under a temperature gradient and become concentrated near the surface as the water evaporates. In such conditions, even when the water table is deep, it is necessary to protect buried concrete against sulphate attack. In addition to the salt-laden capillary water moving past the concrete, there is a risk of the deposited sulphates being taken into solution by rain water or water from shallow sources such as leaking mains or drains.

Recommendations for concrete to be placed in soils and ground waters containing various concentrations of sulphates are given in BRE Digest 250[1] and in BS 8110.[3] The recommendations take into account the type of sulphate which may be present by including in the classifications of severity the amount of sulphate which can be extracted from soils using a limited quantity of water. A saturated solution of gypsum contains 1.2 g SO_3 per litre, so if the sulphate content of ground water exceeds this value, or the sulphate content of an aqueous extract of soil in the proportion of 2:1, water:soil exceeds 1.9 g SO_3 per litre, the presence of the more soluble and more damaging sodium or magnesium sulphates may be suspected.

Sulphates may also be present in materials placed as fill or hardcore. These include demolition rubble containing gypsum plaster, brick rubble, blastfurnace slags and colliery waste shales. Burnt colliery shales have been responsible for most of the recorded cases of attack by sulphates, partly because they often contain high concentrations of the more soluble sulphates, but also because they have been widely used on account of their sound physical properties and ready availability. Unprotected ground floor slabs are particularly at risk of sulphate attack from underlying sulphate-contaminated materials, and considerable damage has resulted. This is illustrated in Fig. 3. The damage generally occurred where the waterproof membrane to the floors took the form of mastic asphalt laid on the upper surface of the slab, contiguous with the damp-proof course in the brickwork of the foundations. Ground water, moving by capillary action under a thermal gradient, was able to carry sulphates from the fill into the concrete, causing sulphate expansion and heaving and cracking of the slabs. The movement of sulphates appeared to be only in a vertical direction. In the many cases of damage to the floor slab, it was found that cement-based mortars in the foundation walls and the concrete footings were completely unaffected. The heave of the slabs did not necessarily take place directly above where the greatest concentrations of sulphates were found in the fill, but in the areas of least constraint, such as in the centre of rooms or in doorways. The heaviest concentrations of sulphates were often to be found under or close to fireplaces where the influence of the thermal gradient was greatest. Diagnosis of the cause of damage to a slab can

Fig. 3. Attack by sulphates on a ground floor slab built on colliery shale

therefore be difficult if the damage is confined to cracking, with no signs of heaving or of displacement of external walls, and when the fill below the cracked areas is found to contain low levels of sulphates.

The burning of colliery waste heaps is thought to be initiated by the heat generated during the oxidation of pyrite in the shales, building up to a level where the residual coal and other combustible materials, such as discarded timbers, are ignited. Old waste heaps are generally poorly consolidated and allow ready ingress of the air and moisture required for the oxidation of pyrite. Sulphuric acid produced during the oxidation reactions dissolves various constituents of the minerals in the shale to give readily soluble sulphates of sodium, potassium, magnesium or manganese, and the less soluble calcium sulphate.

The use of burnt shale for fill was formerly preferred to the use of unburnt shale because it was thought that clay minerals in the latter could swell under damp conditions and cause uplift to buildings. It is now considered that unburnt shale is preferable since it is easier to compact than the burnt shale, its sulphate content is generally lower, and when it is well compacted, access of air is restricted, and with it, the risk of oxidation of pyrite and of combustion. It is stated that damage due to possible swelling of clay minerals can be prevented by placing the shale at close to its optimum moisture content.[4]

CHLORIDES

The chief cause of damage to concrete exposed to external sources of chloride is corrosion of the steel in reinforced concrete. There is a natural risk to structures exposed to sea water or a coastal environment, and even inland structures, at some distance from the sea, can be affected by salt-laden air carried in prevailing winds. The use of chloride de-icing salts on concrete roads has also resulted in damage by frost scaling of the surface or corrosion of reinforcement, particularly on bridge decks.

The action of externally introduced chlorides into reinforced concrete can be more serious than when the chlorides are present in the materials used. For one thing, continual exposure can result in high accumulations of chlorides which can migrate, or diffuse, towards the reinforcement, particularly under conditions of alternate wetting and drying. In addition, the amount of free, and therefore corrosive, chloride, is likely to be greater when chlorides enter hardened concrete than when they are present in the mix, since in the latter case a proportion combines as calcium chloroaluminate during setting. A further factor in corrosion is that chlorides generally do not enter the concrete uniformly in all directions, because of differences in exposure conditions, design details or variations in concrete quality or in depths of cover to the reinforcement. Different chloride concentration cells can therefore occur, and corrosion can be initiated at a number of places on the steel.

Corrosion is defined as the reaction of a metal with its environment, and it can occur by dry or wet reactions. The direct combination of a metal with oxygen may be considered to be dry corrosion and is a process in which metal and oxygen ions react at the metal surface with the uniform destruction of the metal and the simultaneous formation of an oxide layer. The oxides which form on iron differ in chemical composition and in the physical structure of their layers on the metal according to the temperature of the reaction, and range from the simple brown rust formed at ambient temperatures to the complex, three-layered millscale which forms at a high temperature and during subsequent cooling. The oxide layers serve to hinder the passage of oxygen to the metal surface and generally tend to reduce the rate of reaction as the layers increase in thickness.

Wet, or electrochemical, corrosion is a process in which the metals take part in chemical reactions in solution, under the influence of a self-induced electrical current. It is this form of corrosion which affects steel reinforcement in concrete. To understand the process, some knowledge of the principles involved is desirable, and the following brief explanation is given. Conductors of electricity can be divided into two classes – metallic conductors, in which electricity moves without producing chemical changes, and electrolytic conductors, or electrolytes, in which chemical changes always accompany the passage of current. The difference in behaviour is due to the difference between elemental and compound structure. In metals, which are elements, the atoms consist of ions carrying a positive electrical charge and negatively charged bodies called electrons, which balance the ionic charge. Current passed through a metal is carried in one direction only by the transfer of electrons from one atom to the next, while the positively charged particles remain immobile. Solid compounds of the elements, for example sodium chloride, consist of positively charged ions (cations) of sodium and negatively charged ions (anions) of chlorine bonded by their opposing charges. In solution the bonds break down and the compounds no longer exist as such, but as individual ions, each type able to act independently. When an applied current flows through an electrolyte, the ions move in opposite directions, the cations are attracted to the cathode or negative pole, and the anions to the anode, or positive pole. Both types of conductance take part in electrochemical corrosion. The electrical potential which a metal develops in contact with an electrolyte is the stimulus for current to pass, but the corrosive reactions can only take place if oxygen is available at the cathode.

The initial reactions in the electrochemical corrosion of steel are that iron at a site which forms the anode of the cell no longer behaves as a metal, but as ferrous ions and electrons. The electropositive ferrous ions pass into solution in the electrolyte, while the electrons remain in the parent metal and migrate to another site which becomes the cathode of the cell, where they act on the oxygen present in the electrolyte to produce electronegative hydroxyl ions. In chemical terms, the anodic reaction is one of oxidation and the cathodic action one of reduction.

Anodic reaction $\quad Fe \rightarrow Fe^{++} + 2\bar{e}$ (electrons)
Cathodic reaction $\quad O_2 + 2H_2O + 4\bar{e} \rightarrow 4OH^-$

The action in the cell (Fig. 4) is completed by the migration in the electrolyte of ferrous ions towards the cathode and of hydroxyl ions towards the anode. When these ions of opposite charge come together they form a new compound, ferrous hydroxide, which is insoluble in water and is precipitated at some intermediate position between the anode and the cathode of the cell. Unlike the oxide films which form on the surface of the metal during dry corrosion and which slow the rate of corrosion, these insoluble hydroxides form out of contact with the metal and provide no protection. Their movement is restricted by the confines of the anodic and cathodic areas of the steel and it is their accumulation that is primarily responsible for the development of expansive forces capable of fracturing concrete.

In neutral water containing oxygen but not dissolved salts, iron and steel corrode at a relatively slow rate. This can be attributed to the poor conductivity of the electrolyte. When salts are in solution in the water its ionic concentration, and thereby its electrical conductivity, are increased, and the corrosive action is more severe, provided that oxygen is present. The pore water in concrete is a saturated solution of calcium hydroxide containing sodium and potassium salts and is therefore an electrolyte, so it may seem contradictory that under normal circumstances concrete provides a protective environment against the corrosion of embedded steel. This is due to the presence on the metal surface of a passivating film, believed to be either ferrous hydroxide or a lime-rich iron oxide complex, formed by reaction with the highly alkaline pore water which normally has a pH value in excess of 12.5. Even in the presence of oxygen, the film protects the metal against corrosion so long as a high pH value is maintained.[5]

The introduction of chlorides into the concrete promotes corrosion of the reinforcing steel in several ways. Their solution in the pore water increases its conductivity, and hence the activity

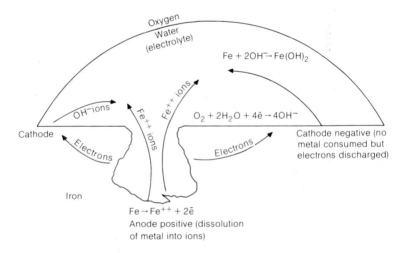

Fig. 4. *Typical corrosion cell*

of the corrosion cell. They can reduce the alkalinity of the pore water to a level which allows breakdown of the protective film and hence allows corrosion to proceed, and they can give rise to concentration cells on the steel. Concentration cells are formed when a single bar is exposed to different concentrations of salts or of oxygen at different points along its length. The areas in contact with the lower concentrations become anodic. If these anodic areas, from which electrons leave the metal, are smaller than the cathodic areas from which the electrons are released into the electrolyte, dissolution of iron in the restricted area of the anode will result in deep pitting of the metal.

A further cause of damage to concrete in which chlorides from external sources may be involved is the alkali–silica aggregate reaction. It is thought[6,7] that alkali chlorides introduced from sea water or de-icing salts may increase the alkali hydroxide concentration of the concrete and so increase the risk of the reaction occurring.

SEA WATER

All waters other than those to be found inland in rivers and lakes are termed sea water, which conveys the impression of a constant chemical composition. It is true that in general the relative proportions of dissolved salts are constant, but their total concentration in the water varies with geographical location. Land-locked seas receiving little influx of fresh water from rivers have higher total concentrations of salts than the oceans. The water of the Mediterranean is nearly 20% stronger in salts than that of the Atlantic, while water in the Arabian Gulf is 25% stronger. Seas receiving large inflows of river waters have lower concentrations of dissolved salts, and areas of the Baltic Sea, for example, contain practically fresh water. The concentrations of the major ions present in the surface waters of the deep oceans are shown in Table 7. In solution, salts are ionised, that is, exist as their constituent ions, which are able to enter into chemical reactions in any combinations, and in the chemical analysis of waters are determined as such. It is customary to express the analytical results in terms of the concentrations of ions present, but it is often easier to evaluate the salinity of a water when the analytical results are calculated to indicate the amounts of salts present. The accepted method of calculating the concentrations of the salts present is based on their solubilities in water, in other words on the order in which they precipitate from solution on evaporation. The analysis of ocean water given in Table 7 shows both the concentrations of constituent ions as determined by chemical analysis and also their theoretical combinations as salts.

The sulphate content of the sea water, expressed as SO_4 in Table 7, is equivalent to 2.3 g/l when expressed as SO_3. This is at a level where precautions would be recommended for the protection of concrete exposed to sulphate-bearing ground waters.[3] Moreover, the action of sea water on concrete is mainly due to magnesium sulphate, and in the ionised state of solution, the sulphate ions can act as if they are entirely combined with magnesium ions, so that the effective content of magnesium sulphate is more than twice the 1.4 g/litre of the theoretical content given in Table 7. However, experience has shown that Portland cement concretes give satisfactory service in marine structures, and this is attributed to the effect that sea water has in inhibiting the expansion that normally accompanies sulphate attack.

The action on concrete of magnesium sulphate in sea water is the same as its action in non-saline solutions. Magnesium sulphate reacts with the calcium hydroxide of the cement paste

Table 7. *Average composition of ocean waters*

Constituent	Concentration, g/l
Ions (determined)	
Calcium (Ca)	0.4
Magnesium (Mg)	1.4
Sodium (Na)	10.6
Sulphate (SO_4)	2.6
Chloride (Cl)	18.6
Combination of ions as salts (theoretical)	
Calcium sulphate ($CaSO_4$)	1.0
Magnesium sulphate ($MgSO_4$)	1.4
Magnesium chloride ($MgCl_2$)	3.0
Sodium chloride (NaCl)	26.8

to give magnesium hydroxide, which forms an insoluble precipitate, and calcium sulphate. It also reacts with the calcium aluminates to give ettringite. This is unstable in magnesium sulphate solution and breaks down to aluminium hydroxide, calcium sulphate and magnesium hydroxide. The role of sea water in inhibiting sulphate expansion has been attributed to the fact that calcium sulphate and calcium sulphoaluminate are more soluble in chloride solutions than in water. However, there are conflicting reports on the behaviour of concrete specimens immersed in sulphate solutions containing chlorides. While some workers have found that expansion is reduced, others have found that it is increased. It seems that other factors are involved when concrete is exposed to sea water, and it has been suggested that because of their increased solubility, the sulphates do not build up within the concrete as they would in static conditions, but are washed out by wave and tidal movements. The expansion which accompanies the formation of ettringite renders the concrete more open to leaching, but in dense concrete, the magnesium hydroxide precipitated during the chemical reactions tends to seal the pores and so reduce the rate of leaching. The action of sea water, therefore, results in a progressive increase in the magnesium oxide content of the concrete and a decrease in the calcium oxide content. In early stages of attack, magnesium oxide levels are highest in the surface layers, but as attack proceeds, the magnesium-enriched zone advances more deeply into the concrete. Sulphate contents tend to rise in early stages, then decrease because of leaching.

Some effects of sea water action are illustrated by an analysis of samples of dense and lean concrete which had been exposed to sea water for 40 years. Analytical figures are of little consequence in themselves when assessing chemical attack and it is useful to reduce them to a basis which allows direct comparisons between samples. In the figures given in Table 8, the basis chosen is the ratio of magnesium oxide (MgO) and sulphate (SO_3) content to calcium oxide (CaO) content, compared with the ratios normal for Portland cements. The figures show that the dense concrete of sample A had suffered very little chemical alteration, while samples B and C had suffered to a greater degree. The lean concrete samples E and F had been more vigorously attacked. Apart from the quality of the concretes, which is always an important factor in the extent to which chemical attack may occur, the dense concrete had been exposed only intermittently to fluctuating water levels, whereas the lean concrete was constantly subjected to full tidal cycles.

Concrete completely immersed below water level is at little risk from chemical attack, as the water can only penetrate to a limited extent and any leaching can only take place by slow processes of diffusion. In the tidal zone between low and high water levels, the attack is more marked, since the concrete is subjected both to leaching and to erosion from wave action and from sands and silts carried on incoming tides. The most vulnerable area of concrete is that immediately above high water level, where the capillary rise of the water and its subsequent evaporation allow an accumulation of salts to take place in the concrete, and these can cause both chemical attack and physical damage by crystallisation. In hot countries, where drying out is possible between high water cycles, concrete in the tidal zone may also suffer disruption from crystallisation of salts. Any spalling, cracking or increase in porosity will allow ingress of chlorides and oxygen to corrode the steel in reinforced concrete.

Table 8. Analysis of samples of dense and lean concrete exposed to sea water for 40 years: ratio of MgO and SO_3 content to CaO content, compared with ratios normal for Portland cement

	Ratio of MgO/CaO content	Ratio of SO_3/CaO content
Dense concrete		
Sample A	0.03	0.03
Sample B	0.07	0.04
Sample C	0.11	0.04
Lean concrete		
Sample D	0.05	0.03
Sample E	0.33	0.17
Sample F	0.59	0.09
Portland cement	0.02	0.04

ACIDS OF NATURAL OCCURRENCE

The acids which occur naturally in soils and ground waters are chiefly organic acids, originating from plant decay, and dissolved carbon dioxide, or carbonic acid. Sulphuric acid, which is inorganic, is included here since under certain conditions it is the product of a natural weathering process.

The action of organic acids and of dissolved carbon dioxide on Portland cement concretes is one of leaching, that is, solution and removal of the cementitious materials, mainly calcium hydroxide, resulting in increased porosity and loss of strength. The action is not expansive as is attack by sulphates, and leaching will only occur to a significant extent if water is able to move through the concrete. Structures subjected to one-sided water pressure, particularly if under a hydrostatic head, are therefore at greatest risk. Structures placed in static water are unlikely to suffer more than surface attack, since once the water in contact with the concrete is saturated with the constituents dissolved from the cement, the reaction ceases.

Carbon dioxide can enter ground or surface waters from the atmosphere, either by direct solution, as in the case of surface waters, or dissolved in rain water, but more abundantly as the result of the decay of vegetable matter. In considering the action of dissolved carbon dioxide a distinction must be made between that which is actively aggressive and that which is unavailable to enter into chemical reactions, and this depends on the nature of the soils or rocks through which the water flows. Where calcium carbonate is present, as in chalks, limestones or marls, the carbon dioxide reacts to form calcium bicarbonate, and although a certain amount of free carbon dioxide remains in the water, it is required to keep the bicarbonate in solution, and is non-aggressive. Uncombined, and therefore aggressive, carbon dioxide is frequently present in significant amounts in waters draining from moorlands where peats overlie siliceous or other rocks incapable of neutralising it. These waters are soft, with a very low content of dissolved mineral salts, but they contain humic acids derived from the peat. The pH values of moorland drainage waters have been reported as between 3.5 and 4.6.[8,9] Water saturated with carbon dioxide at atmospheric pressure has a pH value of 5.7. It has, however, been demonstrated in the laboratory[10] that humic acid is less damaging than carbon dioxide to Portland cement concrete. In this work it was found that a saturated solution of carbon dioxide in distilled water dissolved twice as much calcium oxide from hydrated cement minerals as did a peaty water containing 6% of humic acids, but from which carbon dioxide had been removed. The more aggressive action of carbon dioxide in solution is due to the ready solubility of the calcium bicarbonate which is formed, whereas the calcium salt of humic acid is of very limited solubility. The pH value alone is therefore not a reliable measure of the effects that waters of this type may have on concrete. This is recognised in a number of European standard specifications by the inclusion of the hardness of the water and its content of free carbon dioxide as criteria of aggressiveness, in addition to its pH value.

The effects of soft and moorland waters on concrete in dams have been described by a number of authors, some of whose findings were summarised in a literature review.[11] The majority of the dam faces were constructed using Portland cement, and although leaching was generally evident, particularly at joints and shrinkage cracks, none of the dams was considered to be seriously affected. The degree of leaching was in all cases dependent on the original quality of the concrete. The pH values of the impounded waters were in the range 4.7–7.0, which provides a further illustration of the unreliability of pH values alone to predict chemical action. An interesting feature of the examinations was the finding that while bituminous coatings provided valuable protection to the concrete as long as they remained intact, their useful life was only about seven to 10 years. The dams were constructed before 1955 before synthetic materials, which might have proved more durable, were widely available, but even now, synthetic coatings applied to an area of concrete as large as a dam face could prove very costly, and the provision of good quality, dense and impermeable concrete has been found adequate to provide an acceptable life. Portland cements containing pozzolanic materials are recommended in European standards for use where concrete will be exposed to soft, aggressive waters. For mass concrete construction these cements have the added advantage of generating less heat on hydration than do Portland cements, thus reducing the risk of thermal shrinkage and cracking.

Sulphuric acid is formed under natural conditions when iron sulphide minerals, pyrite for example, undergo oxidative weathering in the presence of air and water by processes which may be both chemical and biochemical.[12] The first stage is a chemical reaction and is the oxidation of the iron disulphide to ferrous sulphate and sulphuric acid. In the presence of certain strains of iron and sulphur utilising bacteria which grow in an acid environment of pH 2.0–4.5, the ferrous sulphate is oxidised to ferric sulphate. A third reaction follows in which the ferric sulphate acts directly to oxidise pyrite and is itself reduced to ferrous sulphate and sulphuric acid, the starting materials for the bacterial reaction. Since the second and third reactions can continue as long as pyrite is available, bacterial activity greatly increases the rate of breakdown of the iron sulphides. It also results in the production of more sulphuric acid than is formed in the chemical oxidation, and this can remain in the free state if the soil contains insufficient carbonate or other basic minerals for its neutralisation.

Pyrite is the most abundant of the iron sulphide minerals and it occurs in rocks of all types and in clays, coals, lignites and peats. Since both air and water are necessary for its oxidation, open deposits of pyritic materials are most prone to develop acidity, and drainage waters from colliery waste shale heaps and coal stacking areas may contain appreciable amounts of dissolved iron salts and sulphuric acid. Under natural conditions, the presence of pyrite in impervious formations or wholly below the water table in unconsolidated deposits does not give rise to acidity. However, disturbance of the natural conditions can induce oxidation and a case has been recorded of the development of strongly acidic conditions, with pH values as low as 1.8, when compressed air was used to keep back ground water during tunnelling in water-bearing sands containing pyrite.[13] The reaction slowed after the air was removed, and within a few weeks the pH value of the water had risen from 1.8 to 3.5.

Damage to buildings, not necessarily associated with chemical attack to concrete, has occurred when oxidation of pyrite in the ground or in fill has been accompanied by expansion, causing uplift of floors and displacement of walls. The pyritic rocks responsible have generally been found to contain calcite, either as a constituent of the rock, as in calcareous shales, or present within fractures in non-calcareous shales. The expansion is due mainly to the action of sulphuric acid on the calcite to produce gypsum. The molecular volume of gypsum is nearly double that of calcite and its formation creates expansive forces. The effects are particularly marked in shales, where the gypsum tends to crystallise within the laminae, forcing them apart. A product of reaction between sulphuric acid and clay minerals is jarosite, a hydrated potassium iron sulphate. This is typically present in weathered shales but its formation contributes little to expansion.

An example of the damage caused by oxidation of pyrite in the ground was given by Penner et al.[14] The basement floor of a building founded on black, pyritic, calcareous shale heaved to an estimated 10.7 cm in six years after construction. It was thought that the ingress into the formation of the air necessary for the oxidation reaction was through an underfloor tile drainage system laid directly on the shale and below a levelling course of crushed limestone. Iron-oxidising bacteria were isolated from samples of the weathered shale. Nixon[15] has reported on widespread damage to houses in north-east England caused by expansive reactions in underfloor fill. The fill material was a locally quarried ironstone containing an estimated 1–2% by volume of pyrite. Although some isolated incidences of floor heave had been observed a few years earlier, the problem did not assume serious proportions until unusually high temperatures were experienced in two successive summers. Microbiological tests were not made during the investigations reported, but it was suggested that the warmer than usual conditions within the fill were favourable for the proliferation and increased activity of iron-oxidising bacteria. Work elsewhere, but unpublished, established that the bacteria were present in shale samples taken from below affected floors and from the quarry face.

INDUSTRIAL CHEMICALS

The number of industrial chemicals injurious to concrete is too great to be dealt with here other than by grouping them into compounds which have the same general composition and behave in the same manner. The actions on concrete of a number of individual substances have been described by Lea[16] and Biczok,[17] and Barry[18] has listed more than 50 chemicals which may be considered potentially aggressive to concrete and has given the manufacturing processes in which they occur. In the construction of new chemical plants or in the repair of damaged concrete in existing plants, the chemicals involved and their concentrations are known and protective measures can be adopted in the areas where the aggressive substances will be handled. A difficult situation occurs where construction is to take place on sites of former chemical industries. Frequently there are no records of what chemicals were used or in which areas of the site. Particularly difficult are filled sites which were used for the disposal of chemical wastes before legislation was introduced to control random tipping and to ensure that records of disposals would be available for the future.

Acidic and alkaline solutions and salts
Acidity and alkalinity are commonly considered in terms of pH values, which, while convenient, are uninformative regarding the type or concentration of substances in solution, and also not applicable to other than fairly dilute solutions. For example, the pH value of a 4.9% solution of sulphuric acid is 0.3. A pH value of 0 will be registered with any increase in concentration up to the 95–98% of the commercial acid. Similarly, at the other end of the scale, a pH value of 14.0 will be given by sodium hydroxide solutions of concentrations from 4.0 to about 50%. The pH scale is logarithmic, and, although not strictly accurate, it can be taken as a rough guide that a tenfold dilution in concentration will raise the pH value of an acid solution by one unit, and reduce the pH value of an alkaline solution by one unit. It can be seen then that the pH values at which concrete is considered to be at risk from acid attack represent very dilute solutions. Nevertheless, no cementitious materials can withstand lengthy exposure to acids, particularly to inorganic acids. It is generally accepted that Portland cement concretes are not resistant to acid conditions of pH values below about 6.0, and that, although other types of cement have greater acid resistance, it is considered that, where pH values are less than 3.5, there is a high risk of damage to concrete made with any type of cement.[19]

The action which organic acids may have on concrete is less predictable than that of the inorganic acids and is more dependent on the solubilities of their calcium salts than on the pH values of their solutions. For example, oxalic acid, which forms an almost insoluble calcium salt, is without effect on Portland cement concretes, whereas acetic acid, which forms a soluble calcium salt, is very damaging. Lactic acid, which occurs in soured milk, is known to attack Portland cement concrete and has caused damage to floors and drains in dairies and butter and cheese-making plants. Butyric acid, which is also a product of rancidity, may be associated with lactic acid, and is believed to have a similar deleterious effect.

Against this evidence, Portland cement concrete tanks have been found to have a reasonable life when used to store fermentation products, containing, among others, butyric, lactic and acetic acids. These acids are present in fodder silage, and the precast concrete staves used in the construction of the silos are generally made with ordinary Portland cement. For durability, it is essential that the concrete is of low permeability. Similarly, tanks made with ordinary Portland cement concrete are used to hold distillation residues containing lactic, acetic and other acids. Although some initial surface attack takes place, the rate of attack decreases with time, and the average, serviceable life of the tanks, before relining is necessary, is said to be 15 years.[20]

Hydrated Portland cements normally have a high pH value, 12.5 or greater, owing to the calcium hydroxide released from the silicates during hydration and to their contents of alkali metal hydroxides. They are therefore relatively immune to the effects of alkaline solutions. However, in the presence of alkaline sodium and potassium solutions, particularly the hydroxides, the risk of damage to concrete from alkali–silica reactions would be greatly increased if the aggregates used were to contain reactive forms of silica.

The chemical action of inorganic salts on Portland cement concrete is confined to relatively few groups of compounds. These are sulphates, acid salts, of which iron and aluminium chlorides are examples, and magnesium and ammonium salts. Physical damage may be caused if strong solutions of non-aggressive salts crystallise within the pores of the concrete, leading to spalling of the surface. Damage may also result from corrosion of the steel in reinforced concrete even though the salts responsible do not enter into chemical reactions with the cement.

The action of magnesium sulphate on Portland cement was discussed earlier in connection with the effects of sea water on concrete. The action of other magnesium salts, chloride or nitrate for example, is more direct, but is also damaging. They react with the calcium hydroxide in the pore water of the cement to form insoluble magnesium hydroxide and calcium salts, chloride or nitrate, to take the examples above. The calcium salts formed are more soluble than calcium hydroxide and are readily lost from the concrete. The overall effect of the magnesium salts therefore is one of accelerated leaching of the cement.

Ammonium salts are particularly destructive to Portland cement concrete. Their action is the same in principle as that of magnesium salts in that ammonium hydroxide and a calcium salt are found. The difference is that ammonium hydroxide is soluble and can also be lost from the concrete. Generally, however, volatile ammonia is evolved. Ammonium sulphate is the most damaging of the ammonium salts, and, indeed, of any sulphates. The calcium sulphate which is formed in the reaction of ammonium sulphate with the calcium hydroxide of the cement is more soluble in ammonium sulphate solution than in water, so attack on the concrete by the sulphate is rapid and severe. Ammonia solution, or ammonium hydroxide, does not attack Portland cements. Ammonia is evolved in the reaction but the calcium hydroxide of the cement pore water remains unaltered.

Sugars

Sugars are widely distributed in the vegetable kingdom where they occur in all parts of the plant – roots, gums, fibre, seeds, sap. Those most commonly found in industrial situations are sucrose, which is the familiar domestic sugar produced from cane and beet, glucose or grape sugar which occurs in fruits and in honey, maltose or malt sugar and lactose or milk sugar. Lactose, while an important consituent of milk and whey, occurs only rarely in vegetable matter. All sugars are powerful retarders of the set and hardening of Portland cements which may be completely inhibited as sugar concentrations approach 1% by weight of the cement. Sugars also attack hardened concrete. They form calcium salts, called saccharates, with the calcium hydroxide of the cement paste, leading to a breakdown of the silicate phases and loss in strength.

Because of their wide occurrence, sugars are to be found in a number of industries. Food manufacturing plants, particularly those processing fruits and where high temperatures may be required, such as in jam-making, present a considerable risk of damage to concrete, especially to floors. In addition to the action of their natural sugars and any which may be added during processing, the organic acids which fruits contain can contribute to chemical attack. All fruits, and many fruit products, such as jam, cider, wine, are acidic with pH values mainly in the range 3–4.

Fats and oils

The animal and vegetable fats and oils which are damaging to concrete are those which are composed of glycerol (glycerine) and a fatty acid. These include the animal fats, such as tallow and lard, and palm, olive and other vegetable oils. Other oils derived from plants, the so-called essential oils, such as turpentine, pine oil and oils used in perfumery, are of different chemical structure, and do not react with cement minerals. The glycerides, that is, the fats and oils of the former group, are saponified by alkalis, a reaction in which a soap is formed from the fatty acids and free glycerol is liberated. In Portland cement concrete, the calcium hydroxide of the cement provides the bulk of the alkali necessary for saponification. The glycerol released also combines with calcium hydroxide to form calcium glycerolate, a compound similar to the calcium salts formed by sugars. There is a gradual breakdown of the silicates of the cement and a weakening of the concrete. The rate of chemical attack is increased if the fats or oils already contain free fatty acids when they come into contact with the concrete. Animal fats, when fresh, do not contain free acids, although on ageing and exposure to the atmosphere they become rancid and acidic. Vegetable oils, however, even when fresh, may contain significant amounts of free fatty acids and are generally more aggressive than the animal fats.

Straight petroleum oils do not react chemically with cements and are without effect on hardened concrete. However, they can affect the hardening of fresh concrete by forming a water-immiscible coating around the cement grains, hindering hydration. Petroleum lubricating and cutting oils which also contain vegetable oils can be damaging to hardened concrete.

Tars and tar oils may be present on sites used for the production of coal gas or metallurgical coke. Although the tars contain phenols and cresols which attack Portland cements, their high viscosity and inability to penetrate hardened concrete render them innocuous. Indeed, coal tars provide excellent protective and waterproofing coatings. The tar oils comprise several fractions of oils distilling over different temperature ranges, and of these, the carbolic and creosote fractions, which are recovered between 210°C and 270°C, are the most aggressive towards concrete. Phenol (carbolic acid) and the cresols, which have similar chemical properties, form sparingly soluble calcium salts with calcium hydroxide, but these do not form a dense protective skin, tending instead to grow outwards from the surface of the concrete in needle-like crystals.

The concentrations at which phenol and the cresols are damaging to matured concrete are uncertain, but the hardening of fresh concrete placed in contact with them can be seriously, even permanently, retarded. It has been reported[21] that when concrete was placed in creosoted wooden sheathing planks covered with a surface coating of cement mortar, neither mortar nor concrete set. The problem was overcome by the use of a coal tar pitch coating on the planks to isolate fresh pours of concrete from the creosote oil.

GASES

Gases in the atmosphere which attack concrete are carbon dioxide, sulphur dioxide, and, indirectly, and under unusual conditions, hydrogen sulphide.

Carbon dioxide occurs in air to the extent of only 0.04% by volume, but is present as a major constituent of exhaust furnace gases. Despite its low concentration in air, it has a significant effect on Portland cement concretes, forming calcium carbonate with the calcium hydroxide present in the cement paste. In dense concrete the reaction is limited to the surface only and can be beneficial in providing a hard protective skin. In porous concretes, carbonation may proceed to greater depths and become a critical factor in the corrosion of steel in reinforced concrete. As discussed earlier, the high alkalinity of the pore water in Portland cement pastes inhibits the corrosion of steel. A saturated solution of calcium hydroxide, which constitutes the bulk of the pore water, has a pH value of 12.4. In cement pastes the influence of the alkali metal hydroxides may increase the pH value to as high as 13.5. A saturated solution of calcium carbonate has a pH value of 9.4. The alkalinity of the pore water is therefore considerably reduced by carbonation. Since there is no gradation of pH between the two values, a simple test can be used to determine the depth to which carbonation may have proceeded. A chemical indicator with a distinct colour change at around pH 9.4 is applied to a freshly broken surface of the concrete. Phenolphthalein is commonly used since its bright pink colour at high pH values is easily seen.

Sulphur dioxide is liberated into the atmosphere from the combustion of sulphur-bearing fossil fuels. It is readily soluble in water and its solutions act as strong acids. A solution of 0.04% has a pH value of 1.5. Where condensation can occur, as in chimneys or railway tunnels, sulphur dioxide emitted from the fuels burnt will dissolve in the condensed water and have a corrosive effect on concrete. In solution as sulphurous acid, sulphur dioxide is slowly oxidised in air to sulphuric acid.

Hydrogen sulphide is generally only present in the atmosphere as the result of the decay of organic matter containing sulphur compounds, and its occurrence is therefore localised. It also occurs naturally in certain volcanic gases and in some spa waters. It is only slightly soluble in water and its solutions behave as weak acids, the pH value of a saturated solution being 4.1. Neither the gas nor its solutions have a marked effect on concrete, but severe corrosion can result when hydrogen sulphide is oxidised to sulphuric acid by bacterial action. This is discussed in the following section. The oxidation to sulphuric acid does not occur in air alone. On exposure to atmospheric oxygen, hydrogen sulphide in aqueous solutions is oxidised to sulphur.

It is often stated that wet chlorine is corrosive to concrete. This is potentially true, since chlorine solutions have the properties of hydrochloric and hypochlorous acids. In practice, however, chlorine gas is so physiologically dangerous that only in the case of accident would it be emitted to the atmosphere in damaging amounts. In the small quantities used for sterilising water supplies, swimming pools, food processing plant, etc., the concentration of the acids produced is very low, and the pH values of the waters remain close to neutral. Chlorine gas is used for sterilisation only in large treatment plants. Other materials, such as bleaching powder, a calcium hypochlorite, and sodium hypochlorite, the bleach of domestic use, are commonly used for the treatment of small volumes of water. These are alkaline in reaction and do not cause damage to concrete.

BIOLOGICAL CAUSES OF ATTACK ON CONCRETE

The biological agents whose activities result in chemical damage to concrete are mainly the micro-organisms – bacteria, fungi and lichens. Although growths of fungi and lichens may appear gross and frequently disfiguring, the individuals which comprise the colonies are of microscopic size and can correctly be considered to be among the micro-organisms. Their activities, however, generally cause only superficial damage. They rarely grow on fresh, uncarbonated concrete but may colonise moist, carbonated concrete which has acquired surface deposits of dirt containing enough organic matter to sustain their growth. Their hyphae, which are analogous to the roots of higher plants, secrete weak acids which can attack the concrete but to a degree only sufficient to allow the organisms to adhere securely. No micro-organisms can exist within concrete, since it presents an essentially hostile environment, being devoid of organic nutrients and commonly of a lethally high pH value. However, the activities of bacteria can cause damage to concrete, not by direct attack, but by the action of chemical products of their metabolism.

The most widely recognised damage of bacterial origin is the destruction of concrete in sewerage systems. Normal domestic sewage is slightly alkaline and does not contain sulphates in sufficient concentrations to attack concrete directly. The damage which occurs is due to sulphuric acid produced by certain bacteria as a result of changes occurring in the sewage. When conditions are such that oxygen in the sewage becomes depleted, its natural and purifying population of aerobic bacteria, that is, those which have an absolute requirement for oxygen, cease to be active. Anaerobic organisms, which do not use atmospheric oxygen but obtain their requirements for growth from salts, reduce the sulphates and various organic sulphur compounds in the sewage to hydrogen sulphide and volatile organic sulphides. These do not directly attack concrete, but on release into the air space dissolve in moisture on the crown of the sewer and the walls above liquid level and are there oxidised to sulphuric acid by aerobic organisms of the *Thiobacillus* species. These belong to the same family as the iron-oxidising bacteria which accelerate the oxidation of iron pyrite to sulphuric acid, and, like them, live in an acid environment and can remain active at very low pH values.

Oxygen depletion, or septicity of the sewage, arises from a number of factors. Low flow velocities, long retention times, particularly in pumping mains and sumps, and the strength, or organic content of the sewage, contribute. Temperature is important in encouraging bacterial growth, so it is not surprising that most cases of attack to sewerage systems have occurred in hot climates. The problem, however, is not unknown in the UK, and it is reported that the number of failures has increased in recent years, owing, it has been suggested, to the move towards fewer and larger sewage treatment works, resulting in longer retention times in the sewers and pumping mains.[22]

A number of measures have been tried to prevent the formation of sulphuric acid, but with varying degrees of success. They include the use of bacteriocides to control microbial growth, increased ventilation or the direct injection of oxygen into the sewage to replace that lost, the addition of lime to raise the pH value of the sewage to a level which inhibits the growth of the sulphur-reducing bacteria, and changes in operating procedures, for example, more frequent pumping cycles or periodical flushing of the mains with water to dislodge the organic slimes and associated bacterial growths from the crowns of the pipes. These are ameliorative measures, usually introduced after septicity has been recognised, but severe damage might already have occurred undetected. The rate of attack can be quite rapid. It has been reported that an asbestos cement sewer outfall in Africa was attacked to the point of collapse within 10 years of installation.[23] The use of limestone aggregate in concrete pipes is said to prolong their life by spreading the acid attack evenly between aggregate and cement paste.[24]

Damage to concrete is also known to have been caused by marine rock-boring molluscs. These live in fairly warm coastal waters and during one stage of their life cycle burrow into the sea-bed rocks. Their role in coastal erosion has been studied extensively but documentation on their effects on concrete is sparse. Biczok[25] has given a reference to reported damage to concrete in the Panama Canal, and Lea[26] has also referred to damage caused by their activities. The molluscs are found in both siliceous and calcareous rocks, where they spend several years, and where, in favourable conditions, they bore rapidly and to considerable depth. A rate of boring of 135 mm per year has been reported[27] but a more conservative estimate is 100 mm per year. The rate of boring varies with species, the hardness of the rock and the conditions of crowding or otherwise under which the animals exist.

The mechanism by which the borers penetrate into rocks is uncertain and has been attributed to both mechanical and chemical means or a combination of the two. It is probable that the species inhabiting siliceous rocks operate in a different manner from those which infest carbonaceous rocks. However, whatever the mechanisms of boring, and regardless of whether the animals can penetrate the cement paste in concrete or only the aggregate, if their burrows extend any distance into reinforced concrete there is a risk that the steel will corrode. Heavy infestation of the concrete by the molluscs could also result in loss of strength, because of the voids created by their burrows.

6
Examination of chemically attacked concrete

Concrete can suffer chemical damage from contaminants introduced with the materials used in its manufacture, by the action of aggressive agents from external sources or by a combination of both. Its examination may therefore appear to present a problem of some complexity, but the likely cause of damage can generally be suggested by the history of the structure.

The first step is to decide whether conditions are such that chemical attack could occur. For this, the factors to be considered include the source of the aggregates used, the location of the affected structure, its exposure conditions and use and which parts have suffered damage. For example, attack on the cement paste by sulphates, or corrosion of steel reinforcement by chlorides, would be considered the most likely causes where marine structures had been attacked. Corrosion by chlorides may also be suspected in other situations, for example, where de-icing salts are known to have been used on concrete road surfaces or where the concrete is of such an age that calcium chloride might have been used as a set accelerator. The use of aggregates containing harmful amounts of sulphates or chlorides may be suspected as a cause of damage to concrete made in countries where it is known that available aggregates are frequently contaminated with these salts; or the aggregates used may belong to the rock types likely to enter into alkali–silica reactions. A concrete floor slab laid in non-industrial premises may have suffered attack from sulphates from fill material placed below it, whereas in premises used for manufacturing, a floor may be damaged from the surface by chemical spillages. Similarly, the underside of a concrete wharf is most at risk of damage by sea water, but the surface may suffer attack from contact with materials regularly handled, for example, fertilisers containing ammonium sulphate or nitrate.

The second step is close visual inspection of the concrete. Observations of patterns of cracking, seepages, discoloration, staining, spalling, or softening or erosion of the surface can give valuable indications of the possible cause of damage. However, caution is required before conclusions are reached, since the same physical effects may be produced by different chemical reactions. Cracks on the line of the reinforcement are a sign of corrosion of the steel, but cracking caused by the expansive alkali – silica reaction of aggregates also tends to follow the line of the steel in reinforced concrete. Where corrosion has occurred there may be rust staining in or near cracks, but absence of staining does not necessarily mean that corrosion has not occurred. Conversely, rust stains on the surface may be unrelated to corrosion of reinforcement, and may result, for example, from the oxidation of reactive forms of iron pyrite in the aggregate.

The third step is for the concrete to be examined in the laboratory, and for this to be informative, samples must be selected carefully. If it is required only to establish whether the concrete has been subjected to chemical action, the most suitable sample is often one taken from an area where the greatest damage is apparent. However, this is not always the case, since the seat of chemical attack may be remote from where its effects are seen. As a general rule anything unusual should be sampled, such as cracked or discoloured areas. The analysis of surface deposits or of seepages through the concrete is generally not very useful, since these always contain calcium hydroxide and sulphate, and possibly alkali metal salts leached from the cement. In some circumstances, where there is reason to suspect that concrete has been attacked by a chemical substance which is not a normal constituent of cement – ammonium nitrate, for example – tests showing the presence of nitrate in seepages will provide confirmatory evidence. Samples of concrete should be taken to some depth, and, wherever possible, a core should be cut.

When the cause of damage is known and it is required to establish the extent of chemical reactions within the concrete, it is necessary to institute a representative programme of sampling, and for this, it is convenient, and often more informative, to take powdered samples using a masonry drill. This is recommended in BRE Information Sheet IS 13/77[1] as a preferred method for sampling building structures to determine the distribution of chlorides throughout the concrete. The technique required to obtain samples suitable for chemical analysis is described, and guidance is given on the frequency of sampling and appropriate locations. Drilling can also be useful in the laboratory to take samples at various distances along cores which have been cut for physical tests on the concrete. Samples from concrete components made with, or suspected of having been made with, high alumina cement are normally obtained by drilling. Only small amounts of material are needed for the determination of chloride or for tests which may be made on high alumina cement concrete, and drilling yields an adequate quantity. It might be thought that samples obtained in this way would be unrepresentative of the concrete and that there would be a bias towards a greater proportion of cement paste or of aggregate, whichever is the softer, but, in fact, they are remarkably consistent.

In the laboratory it is frequently worth while to make a preliminary microscopical examination of a sawn and ground sample of concrete. This can show the presence in voids or in cracks of secondary crystalline growths, such as ettringite formed by attack caused by sulphates, or of gel exudations produced in the alkali–silica reaction. Observations of the size, number and shape of voids and of micro-cracking within and around aggregate particles can give supplementary information on the condition of the concrete when it was placed and the influence of moisture movements of the aggregates. Deleterious impurities, such as clays, soft materials or coal, which generally originate in the aggregates, can be recognised and their content, if required, determined by other tests.

Where the instrumentation is available, analysis by X-ray diffraction can also be valuable in deciding whether certain forms of chemical attack have been the cause of damage to concrete. Although it is an informative technique, in the examination of concrete it cannot be considered to be other than qualitative or semi-quantitative, since it is necessary to separate the cement matrix from the bulk of the aggregate. However, ettringite and the low-sulphate form of calcium sulphoaluminate can be determined with a fair degree of accuracy by comparing their contents with those of other cement minerals. Examination by X-ray diffraction is also useful to distinguish different types of cement, and the degree of hydration of Portland cements. Some of the more common forms of chemical attack on concrete are discussed in the following sections.

ATTACK BY SULPHATES AND SEA WATER

Visual examination of the concrete can give a useful guide on whether attack by sulphates should be suspected. Attack is accompanied by expansion and frequently by cracking, although there is not a distinctive pattern of cracks. Where the concrete has heaved or bulged, the cracks often radiate from the centre of movement. In other cases the pattern is random. Expansive cracking may be the only sign when the attack is caused by sulphates present in the aggregates. Concrete attacked from external sources often exhibits a whitened appearance, more apparent on drying. This is indicative, but not conclusive, of action by sulphates, since the same effect can result when water has passed through concrete which has one side exposed to the atmosphere. Calcium hydroxide, leached from the cement paste, carbonates on exposure to the air, leaving a white deposit of calcium carbonate on the concrete. The action of sodium or calcium sulphates leaves the concrete soft and mushy while damp, although some hardening may occur on drying. Attack by magnesium sulphate is more difficult to diagnose by visual means, since the surface which has been attacked is hardened by the deposition of insoluble magnesium hydroxide within the pores of the cement paste.

Damage caused by sulphates frequently results from their action on concrete floor slabs made with ordinary Portland cement and laid without the provision of a waterproof membrane below them. Sulphates from the ground, or, more usually, from contaminated fill, are carried upwards in capillary water moving under a thermal gradient. The movement is generally vertical in direction, and concrete in footings and cement–sand mortars in brickwork foundations are normally unaffected. Strip foundations acting as retaining walls on sloping sites can be damaged by sulphates in water flowing down the site. Heaving of the slabs is usually the first sign of attack by sulphates, often first noticed as difficulty in closing doors, since expansion mainly occurs in areas of the floor under least restraint. There may also be lifting of partition walls, accompanied by cracking. Outward displacement of external walls at or just above damp-proof course level may follow, and the induced stresses can result in cracking of the superstructure. In thick slabs the expansive movement may be in a horizontal direction only, and displacement of the walls of the building may be unaccompanied by heaving.

Heaving of floors is an example of an instance where sampling the concrete and associated materials only in the area of greatest apparent damage can lead to inconclusive results concerning the cause. Although water generally moves vertically in the cold sub-grade, in the warmer conditions immediately beneath the concrete slab, it often moves laterally towards the warmest areas of the floor. Higher concentrations may therefore be found in these areas of fill than immediately below where the damage is manifest. A damaged floor should be sampled in several positions through the full depth of the concrete, and samples should also be taken of fill and soil below. The determination of the total sulphate content of sub-samples representing the underside and the surface of the slab is usually sufficient to establish whether attack by sulphates has occurred.

When concrete has been exposed to the action of sea water, samples for analysis should be taken at several depths within it in order to establish the extent to which chemical attack has occurred. As described in the previous chapter, in concrete attacked by sea water the content of calcium hydroxide progressively falls as a result of leaching, while the content of insoluble magnesium hydroxide deposited in the cement paste increases. Calcium sulphate is also removed by leaching and the content of sulphate in the concrete tends first to rise and then to fall. Comparison of the ratios of the calcium and magnesium contents will indicate the depth to which reaction has proceeded, but more information is generally required, such as the extent to which chlorides have entered reinforced concrete, and the contents of total sulphate and of sulphate which has combined as ettringite. Where concrete has been attacked by non-saline waters containing magnesium sulphate, the soluble reaction products are frequently not removed by leaching, and the determination of total sulphate and magnesium oxide contents is all that can usefully be done.

The extent to which concrete may have suffered sulphate attack is generally assessed by expressing the sulphate present as a percentage by weight of the cement content. However, if the cement has been significantly altered by chemical reactions that have occurred, its original content in the concrete cannot be determined reliably. In severe cases of attack, the sulphate content in the concrete may approach or exceed the amount to be expected in the cement alone. Lower concentrations may not indicate attack so clearly unless the sulphate found can be related to the cement content. This can be taken to be the amount specified in the mix, or, if this is not known, an inverse form of calculation can be applied to the test results. In this case it is assumed that all the sulphate is derived from the cement, and a cement content is calculated accordingly. The engineer can then decide from other considerations, such as the quality of the concrete and where it had been placed, whether the calculated cement content was a realistic value, or whether the sulphate content in the concrete was excessive.

ATTACK BY CHLORIDES

Plain concrete is not chemically affected by chlorides unless they are present in very high concentrations such as may be used in manufacturing industries. The chief cause of damage is the corrosion of steel in reinforced concrete, leading to cracking, and, in severe cases of attack, spalling of the concrete cover. Cracking is typically in the form of fairly straight lines following the reinforcement, as shown in Fig. 5. Rust stains close to cracks indicate corrosion, but not necessarily by chlorides, and not necessarily as the primary cause of damage. Once cracks open in the concrete, for whatever reason, which may be purely physical, the steel is at increased risk of corrosion, even from the atmosphere. There will be more ready ingress of oxygen and moisture, and carbonation of the cement paste will be increased within the cracks, reducing its protective alkalinity. Corrosion by chlorides may be suspected if the concrete has been exposed to a marine atmosphere or if there is reason to believe that contaminated aggregates might have been used.

The total chloride content of concrete is determined in the laboratory by a straightforward test. Test methods which can be used in the field are given in BRE IS 13/77 and BRE IS 12/77.[2] The methods are based on the use of commercially available water-testing kits and although they are not absolutely accurate they are useful for comparing chloride levels at different points within a structure and for indicating areas where further samples should be taken for analysis. Since chlorides are generally unevenly distributed within concrete, the field tests could also have application in the repair of damaged concrete, when the extent of breaking out is decided on the basis of a predetermined level of chloride.

It is considered that the total chloride content of the concrete, although readily determined, does not correctly indicate the corrosion risk to embedded steel. The total chloride content is determined after extraction of the concrete with nitric acid and includes the chlorides combined in the aluminate and ferrite phases and bound in the calcium silicate hydrates as well as the free chloride in solution in the pore water. As earlier discussed, a small amount of chloride is required to stabilise the

Fig. 5. Cracking of concrete caused by chloride-accelerated corrosion of reinforcing steel

Examination of chemically attacked concrete

chloroaluminates and ferrites formed when chlorides are present during hydration of the cement. It is this free chloride, and, in the case of contamination from external sources, any introduced into the hardened concrete, which promotes corrosion. However, the free chloride is not easily determined. Extraction of the concrete with water results in the release of part of the combined chlorides, and, indeed, the whole, or very nearly the whole, of the total chloride present can be recovered by successive extractions with water. One research method measures the free chloride in the pore water by extracting the liquid phase of hardened concrete under high pressure.[3]

A more practical approach was adopted to determine the amount of concrete to be broken out of a bridge deck damaged by corrosion of the reinforcement caused by the use of de-icing salts. In this method, the free chloride was estimated by extracting a sample of concrete with different volumes of water and extrapolating the results to the moisture content of the concrete. The ratio of free chloride so estimated to the total chloride content found by analysis was used to establish a total chloride threshold level above which the concrete was required to be replaced.[4]

Non-destructive test methods have proved valuable in determining the condition of the steel reinforcement in concrete before there are visible signs of corrosion. Methods based on electrochemical principles were developed in the USA in the late 1950s when widespread damage had been caused to reinforced concrete bridge decks treated with de-icing salts. It was realised that, to examine the great number of bridges at risk from the same cause, there was a need for an inspection method to detect the onset of corrosion without having to break open the concrete. The method developed uses a half-cell probe, generally copper/saturated copper sulphate solution, connected through a high impedance millivolt meter to the steel reinforcement. The base of the half-cell is placed on the surface of the concrete in a regular grid pattern and the readings of the potentials developed are plotted as a map showing lines of equal potential. A considerable amount of work was done by Stratfull[5,6] to correlate the half-cell potentials measured on a bridge with visual observations of the extent of corrosion of the reinforcement. His work established ranges of potential given by actively corroding steel and those given where the steel had not corroded. Bars in which no corrosion was observed were associated with potentials more positive than −320 mV, measured with the copper/copper sulphate half-cell, while corroded bars gave readings more negative than −350 mV.

A further development of the technique was the use of two half-cells applied to the surface of the concrete, one in a fixed position remote from the line of the reinforcement, the other moved to give a grid pattern of readings across the surface.[7] This technique has the advantages that no connection to the steel is

Fig. 6. Damage to concrete caused by alkali−silica reaction

required and that any uncertainty about the electrical continuity of the reinforcing bars is removed. Although it does not give the actual electrical potentials of the steel, it identifies anodic and cathodic areas. Positive equal potentials indicate anodic areas and therefore corrosion of the steel, while negative equal potentials indicate cathodic areas. The readings of potential do not provide information on the rate of corrosion, but an indication of the extent and severity can be obtained from the patterns of the lines of equal potential given by the single cell probe or from the contours of comparative potential given by the two cell probes.

Half-cell measurements have been useful in indicating corrosion of reinforcing steel in concrete, but the test method is not infallible. The potentials recorded are sensitive to differences in the moisture content of the concrete in different areas of a structure, so that any sudden change in potential readings might not indicate the onset of corrosion, but merely a difference in moisture content. Where the readings do indicate corrosion, it is advisable that confirmatory chemical tests should be made on the concrete and the steel.

ALKALI–SILICA REACTION

The reaction between alkaline hydroxides in the pore water of the cement paste and reactive forms of silica in the aggregates used is expansive, and results in cracking of the concrete (Fig. 6). In unrestrained concrete, the cracks have a characteristic linked pattern, described by early workers on the problem as map cracking. A typical pattern is shown in Fig. 7. Where there is any restraint on the expansion, as in reinforced concrete, the cracks tend to be linear and parallel to the reinforcement. In some cases (Fig. 8) there may be exudations through the cracks of a colourless, viscous gel composed of sodium, potassium and calcium silicates, which carbonate on exposure to the air, becoming hard and white. The calcium hydroxide which normally weeps from cracked concrete also carbonates in air, so that any white deposits which may be observed are not proof of the alkali–silica reaction. Damp patches, the appearance of gels around aggregate particles near the surface and surface popouts are also indications of the reaction. Chemical analysis of surface deposits is not very informative, because of the influence of the constituents which may be leached from the cement. It is stated that microscopical examination of dried white gel deposits on the surface of the concrete can distinguish the gel from crystalline calcium hydroxide and other hydration products of the cement,[8] but the only positive proof that alkali – silica reaction has occurred comes from microscopical examination of the interior of the concrete. The reaction is identified by the presence of gel in voids and cracks, and by rims of alteration products around reactive aggregate particles. Gel deposits are usually found in close proximity to the reactive particles. The microscopical examination may be made of a sawn and polished surface at a magnification of about × 50.[9] A thin section is recommended in BRE Digest 258.[10] A test to confirm that the alkali – silica reaction has occurred in concrete and also to indicate the potential for continuing expansion is given in

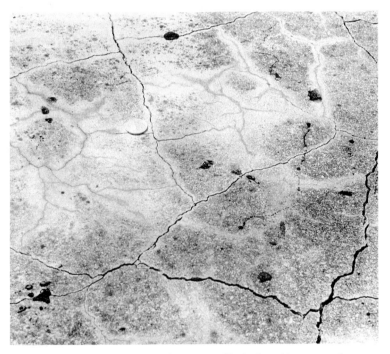

Fig. 7. Alkali – silica reactions in concrete. Typical cracking pattern in unrestrained concrete

Digest 258. It is the measurement of length changes of cores stored at elevated temperatures and 100% relative humidity.

The development of the alkali–silica reaction to the extent where cracking is visible is fairly slow, generally about five to six

Fig. 8. Alkali–silica reactions in concrete. Exudations of reaction products through cracks

years, although in some instances it may be quite rapid. The age of the concrete when cracking is first observed may therefore suggest whether chemical reactions or physical effects, such as drying shrinkage, were responsible.

ACTION OF ACIDS

The action on Portland cement concretes of soft waters or acid waters containing free carbon dioxide or humic acids is to leach calcium hydroxide from the cement paste. In dense concrete, the exposed face often has an etched sandy appearance, with the coarser particles of the fine aggregate fraction standing proud of the surface. Leaching is generally more pronounced near cracks and joints or in any areas of honeycombing. Where water can pass through concrete from one side there may be white deposits of lime or of calcium carbonate on the other face. Water containing appreciable amounts of humic acids often leaves brown stains on the concrete.

In general, chemical tests on leached concrete do not provide more information than can be obtained by other means, such as measuring the depth to which the concrete surface has been softened, or in the case of exposed structures has been eroded after softening. There are occasions, however, when it is required to distinguish between leaching and other forms of damage – frost action, for example. Although acid waters remove calcium hydroxide from the cement paste, they often leave appreciable amounts of their corresponding calcium salts in the concrete. Whether acid leaching has occurred and to what extent can be found by comparing the ratios of soluble silica and of calcium oxide in the cement–sand mortar of the concrete with those normal for Portland cements, and by determining the acid radicals which may be present, carbonate for example, where the damage is thought to be due to the action of carbon dioxide. Microscopical examination can also be informative, as in the pioneering work of Terzaghi,[11] who diagnosed the cause of damage to concrete submerged in sea water as being due to the action of aggressive carbon dioxide. She compared the amounts of crystalline calcium hydroxide (Portlandite) and calcium carbonate (calcite) in the cement paste of deteriorated concrete samples with those present in samples taken from an exposed and unaffected part of the structure.

Examination of chemically attacked concrete

While attack by acids can be considered in general terms to be a process of leaching, in which calcium hydroxide and other soluble constituents of the cement paste are removed, resulting in its disintegration, the action of sulphuric acid is also accompanied by expansive attack by sulphates. When the acid first comes into contact with the concrete it reacts with the calcium hydroxide of the cement to form calcium sulphate, or gypsum, which then attacks the hydrated calcium aluminates to give ettringite. This is only stable at a high pH value, and as the acid penetrates further into the concrete, lowering the pH, ettringite breaks down to gypsum and hydrated calcium and aluminium compounds. Attack by sulphuric acid can be distinguished from that caused by neutral sulphates by chemical tests to determine whether the sulphate is combined as gypsum or as ettringite.

HIGH ALUMINA CEMENT CONCRETE

Although the use of high alumina cement in structural concrete was discontinued some years ago, a number of buildings still incorporate units made with it, and investigations still continue into their soundness and the extent to which the cement may have suffered conversion. Characteristically, concrete which has suffered conversion has a brown appearance, more noticeable within the concrete than on the surface. The change in colour is due to the oxidation of ferrous iron compounds in the cement, and strictly is more an indication of porosity than of conversion. It can be found by qualitative chemical tests whether concrete has been made with high alumina cement. The degree of conversion is measured by the instrumental technique of differential thermal analysis. This is based on the principle that any changes in the structure of a substance when it is heated will be accompanied by either absorption or emission of heat. The material under test is heated at a steady rate alongside a thermally stable reference material, generally calcined aluminium oxide. The temperature at which the test substance emits or absorbs heat is indicative of the reactions it has undergone and can be used to identify particular groups of compound. In testing high alumina cement concrete, the compounds present in unconverted cement and those which form during conversion are identified and the degree of conversion found by comparing the amounts of each present.

7
Specifications for chemical testing

Standard specifications which cover the chemical properties of materials generally include or stipulate the test methods to be used. It may seem, therefore, that the engineer will have no difficulty in ordering chemical tests by specification number and test number, and although this may safely be done in many cases, in others the method selected may be quite inapplicable to the material to be tested. Particularly in problem areas, such as in assessing the risk to concrete or in establishing the cause and extent of any chemical damage concrete has suffered, and frequently in less problematical fields, it is preferable to discuss the test requirements with the analyst, who can then decide what methods are most appropriate to the sample, to the information required from the analysis and to the use to which the results will be put.

It is not always appreciated that, with few exceptions, chemical test methods were not devised specifically for inclusion in standard specificatons, but had been in use for many years, centuries even, before the specifications were written. Specification test procedures are mainly wet methods of analysis, in which the constituents are determined in solution. They are methods of proven reliability that can be followed without the need for sophisticated apparatus. However, most chemical constituents can be determined by more than one method of analysis, and the introduction of instrumental techniques, such as flame emission photometry, aborption spectophotometry and X-ray fluorescence spectrometry, has greatly increased the choice available to the analyst. This is recognised in a number of British Standard specifications by the qualification that a method of equal accuracy may be used instead of the one given.

Specifications for chemical testing

There are, however, circumstances where only the most rigid adherence to standard procedures will yield meaningful and comparable results. Certain ignition tests and the preparation of aqueous extracts are examples. The use of such instrumental techniques as X-ray diffraction, microprobe analysis or electron microscopy can be of great value in certain abstruse aspects of the examination of concrete or its constituents, but generally their application is more suitable to research work than to the practical problems of the chemical reactions that can occur in concrete.

TEST METHODS FOR SOILS, GROUND WATERS AND FILLS

The test methods most relevant to the performance of concrete to be placed in natural soils are generally confined to the determination of the total sulphate contents and the pH values of soils and ground waters, and, when soil sulphate contents are high, to the determination of water-soluble sulphate. Methods for these tests are given in BS 1377[1] and in BRE Current Paper CP2/79.[2] The latter also includes methods for the determination of chloride, sulphide and ammonia and other cations. These tests are not routinely required for the examination of natural soils, although sulphides may be present where conditions are anaerobic, as on waterlogged sites. If conditions are altered to allow free ingress of air, by drainage for example, the sulphides can become oxidised to equivalent amounts of sulphates. Chlorides may be present as a result of sea-water infiltration, or may occur in significant amounts in soils and ground waters in arid regions.

The total sulphate content of a soil is determined in a hydrochloric acid extract, which takes into solution any calcium sulphate, gypsum, that is present as well as the water soluble sodium and magnesium sulphates. Gypsum is of limited solubility in water, 1.2 g per litre as SO_3, so although its content in a soil may be high, its rate of solution and hence its degree of attack on concrete will be less than that of the more soluble sulphates. The presence of readily soluble sulphates is indicated by extracting the soil with a limited and carefully controlled amount of water. If gypsum is the only sulphate salt present, the maximum amount that will be dissolved when 100 g of soil are treated with 100 ml of water will be 0.12 g of SO_3 or 1.2 g of SO_3 per litre of extract. The test method given in BS 1377 is for the extract to be prepared using a ratio of 1:1, soil:water. The site classifications of aggressiveness given in BRE Digest 250[3] and in BS 8110[4] are based on an extraction ratio of 1:2. soil:water. Table 2 of Digest 250 correlates the sulphate values obtained by the different ratios of soil to water, and its recommendations for precautionary measures can therefore be followed if the tests have been made using the BS 1377 method of extraction. In practical terms, the higher ratio of water is to be preferred since an adequate volume of extract for analysis is more easily obtained, particularly from clayey soils. There is no benefit in specifying that tests should be made using both ratios. Different results are often obtained, in which case it is recommended in Digest 250 that the site classification of aggressiveness should be based on the sulphate contents of the 1:2, soil:water extract.

The measurement of the pH value of soils or waters is usually made with a pH meter. Colorimetric methods that use mixed indicator solutions are also applicable and a supplementary colorimetric method is included in BS 1377. This was intended primarily for site use, but has been widely replaced by the use of portable pH meters, which are now readily available, inexpensive and more convenient to use. Robust types of electrode are also available, and these should be selected for site work. The pH values indicate whether a soil or water is acidic (pH value less than 7.0) but give no information on which acids are present, and there are no standard specification tests that will do so. In natural conditions acidity is most likely to be due to the presence of humic acids or of sulphuric acid arising from the oxidation of pyrite or other iron sulphide minerals, or, in ground waters, of free carbon dioxide. A very low pH value, of less than about 3.5, will suggest that sulphuric acid is present, and although it might be thought that this would be confirmed by a high sulphate content, that is not necessarily the case. A solution of sulphuric acid having a pH value of about 3 contains only 0.04 g per litre of SO_3. Iron pyrite is not attacked by hydrochloric acid and it will not contribute to the total sulphate content determined by the standard test procedure. In the absence of sulphides that are readily decomposed by hydrochloric acid with the evolution of hydrogen sulphide, the

Specifications for chemical testing

presence of pyrite may be suspected if the total sulphur content of the soil, when expressed as SO_3, is greater than that of the hydrochloric acid soluble sulphate.

The problem of specifying chemical tests on the sites of former chemical manufacturing industries or on those that have been filled with waste materials is one of considerable complexity. It may be known that only one process has been carried out on the site, and then the tests required to assess the risks to concrete can be decided with some confidence. A more difficult situation occurs if the site has been occupied by a succession of manufacturers or has been used for the production of a number of different chemicals. Frequently only incomplete information is available about the processes, the areas in which they were carried out and where waste products may have been dumped on the site. Even when these facts are known, site clearance and levelling may have resulted in a spread of contaminants across the entire area. Also difficult are filled sites in which industrial wastes may have been deposited, and even those which ostensibly have received only domestic refuse may contain unsuspected pockets of chemical contamination.

Possibly the best approach when dealing with filled ground is to have chemical tests made to identify all unusual substances encountered, such as crystalline deposits and materials having a pronounced colour or smell. Not all chemical wastes are deleterious to concrete, and, in the first instance, measurements of pH value and qualitative tests will indicate the need or otherwise for detailed quantitative analysis. It is fortunate that recent concern about the health hazards of filled and industrial sites has led to the adoption of more comprehensive sampling programmes than might be required for engineering purposes, and hence the number of samples recovered and available for analysis is less limited than formerly.

Materials used as hardcore may contain sulphates in amounts sufficient to cause chemical attack on concrete placed in contact with them. The types most likely to cause damage are colliery waste shales, blastfurnace slag and concrete and brick rubble. Concrete rubble alone is not deleterious, but when it comes from building demolitions it may contain gypsum plaster, which can cause attack by sulphates in wet conditions. Similarly, brick rubble from demolition often has adherent gypsum plaster,

while new bricks may have a naturally high content of sulphate.
The types of material that are used as hardcore, and their physical and chemical properties, are described in BRE Digest 276[5] and BS 6543.[6] The chemical property that is considered likely to cause damage is the sulphate content of the fill, and it is recommended in both documents that the risk of attack should be assessed from the sulphate contents of 1:2, fill:water extracts, rather than on the total acid soluble contents. A method for the preparation of aqueous extracts of blastfurnace slag is given in BS 1047,[7] which is a specification for graded coarse aggregates of up to 40 mm. The sulphate determinations are made on extracts of two prepared fractions smaller than 20 mm, the material up to 40 mm being crushed to pass the 20 mm sieve. A considerable range of particle sizes can be expected in slag supplied for use as hardcore, and care must be taken to ensure that the crushed fractions are blended to reflect the proportions of coarse and fine material in the slag as sampled. The reason for the use of a fairly coarse grading for the preparation of aqueous extracts of slag is that the sulphur compounds that it contains may be bound within vitreous or relatively impermeable particles, and may therefore be inaccessible for solution in water. In more permeable types of hardcore, such as bricks, or shales, which have a laminar structure allowing ingress of water, it is preferable that the aqueous extracts should be prepared from crushed and finely ground samples.

TEST METHODS FOR CEMENTS

Methods for the chemical analysis of cements are given in BS 4550:Part 2[8] which includes tests recommended by the International Standards Organisation. There are some differences between the two sets of methods and those recommended for use in the UK are indicated. The main difference is in the determination of aluminium oxide, which in the UK test method is analysed directly and more precisely than in the International Standards method. The test methods were designed to give accurate analyses of all types of Portland cement, but are not entirely applicable in all respects to the analysis of other cements, such as high alumina cement or blastfurnace slag cements. They are also not fully applicable to Portland cements containing pozzolanic materials. In addition to a general

Specifications for chemical testing

scheme of analysis to establish compliance with specification requirements, BS 4550 gives test methods for the determination of various minor constituents and for the determination of the alkali metals, sodium and potassium. Although limited in only a few national specifications for cements, their content is important when the aggregate to be used is of a type that may be suspected of entering into alkali–silica reactions. Also included is a chemical test for pozzolanicity, which certain pozzolanic cements are required to satisfy.

The only pozzolana used to date in the UK is pulverised fuel ash. The required properties of ash to be used in structural concrete are given in BS 3892:Part 1.[9] This includes a test method for the determination of magnesium oxide content by atomic absorption spectrophotometry. The methods given in BS 4550 are for the determination of sulphate content and loss on ignition of the ash. The same test methods are specified in British Standards for Portland pulverised fuel ash cement[10] and for pozzolanic cement with pulverised fuel ash as pozzolana.[11] Both these specifications include a method for the determination of the proportion of pulverised fuel ash in the cement.

TEST METHODS FOR AGGREGATES

Tests routinely required on natural aggregates are the determination of sulphate and chloride contents, and, for marine-dredged materials, shell content. Tests for the potential reactivity of certain types of aggregate with alkali metal hydroxides in the cement have been made for many years on overseas aggregates, but it is only in recent years that the deleterious reactions have occurred in the UK, and, with them, a need for the tests to be made on local aggregates.

Although standard specifications governing the properties of aggregates place limits on their sulphate, chloride and shell contents, test methods for the determinations are not clearly defined. In particular, there is a lack of guidance on the manner in which different materials should be prepared for analysis. For example, in the test method for the determination of chloride content given in BS 812,[12] the chlorides are extracted by water without any need for the samples to be crushed. While this procedure is suitable where chlorides are present on the surface only, as in the marine-dredged quartz sands and flint gravels which are the most likely of UK aggregates to be contaminated, the procedure is not to be recommended for testing crushed limestone and other carbonate rocks of evaporite origin which are used as concreting aggregates in other parts of the world and in which the chloride is distributed throughout the particles. These materials need to be crushed and finely ground before the chloride can be completely extracted by water, and even then the method of BS 812, in which large sample weights are used and the analysis is made on aliquots of the water extract, is not the most convenient to use with fine materials. A more practical approach is to use a small sub-sample, about 10 g, of the ground material, and to extract it with either water or nitric acid and use the whole of the extract for analysis.

With the exception of blastfurnace slag dense aggregate, for which test methods are detailed in BS 1047, the sulphate contents of aggregates are required to be determined by the method given in BS 3681,[13] in which the test is made on a finely ground sample. While it is necessary to treat most types of aggregate in this manner, an exception could be made in the case of flint aggregates, where any sulphates will be present only on the surface. For these materials an approach similar to that specified for extracting chloride could be adopted with some modifications. The extraction should be made with dilute hydrochloric acid and precautions should be taken to prevent excessive effervescence due to the solution of shells or chalk particles.

The determination of shell content also is not well covered by specification methods. A direct method, given in BS 812:Part 106,[14] consists of separating the shells by hand and weighing them, but applies only to coarse aggregate of greater than 10 mm particle size. Other test methods do not measure the shells directly, but rather the carbonate or acid-soluble material, and therefore include any chalk and other soluble substances that might be present. Although no limits are given in BS 882[15] for the shell content of aggregate fractions finer than 5 mm, tests are often requested on sands, and a suitable method is given in BS 812:Part 119.[16] The test measures acid-soluble material and was intended specifically for testing fine aggregate to be used in concrete for road surfaces or pavement blocks. The standard procedure is for the tests to be made on fractions of the aggregate coarser and finer than 600 μm. For concrete for other

Specifications for chemical testing

purposes, the sample can be tested as a whole. Limits are placed on the shell content of 5–10 mm aggregates, but there is not a test method specific to this size fraction. A combination of both other test methods is probably the best approach, the obvious shells or shell fragments being removed by hand and the smaller ones being dissolved in acid.

The tests that have most generally been used to determine whether aggregates containing certain forms of silica will react expansively with cement alkalis were developed in the USA and have been adopted as ASTM standards. They are petrographic examination,[17] a chemical test method[18] and the measurement of length changes of mortar bars.[19] Another test that has found some use, although it is not included in standard specification methods, is the gel pat test. While each test has its merits, no single test can accurately predict the behaviour of an aggregate in practice. This is emphasised in an appendix to ASTM C33[20] in which it is stated that the various methods proposed for evaluating the potential reactivity of an aggregate do not provide quantitative information on the degree of reactivity to be expected or tolerated in service, and that assessment of the risks should be based both on test data and on the examination of concrete structures which had been made with the fine and coarse aggregates and the cement proposed for use in the new works.

Of the ASTM tests, petrographic examination of the aggregate provides positive identification of the presence and amounts of the silica minerals that are known to react with alkalis, but it does not predict the expansion that might result when the aggregates are used in concrete. Nevertheless it is a useful method for assessing potential alkali reactivity and it can be carried out fairly quickly. Measurements of the expansion of mortar bars made with a high alkali content by weight of cement have been considered to be the definitive method of determining alkali reactivity and are recommended in the ASTM standards for confirming potential reactivity indicated by other tests. The mortar bar test, however, is lengthy and significant expansions may not be recorded earlier than 12 months with slowly reacting aggregates. The specification gives limits for expansions at 3 months and 6 months which indicate reactivity. The time taken to obtain results from mortar bar expansion measurements may be acceptable when new aggregate sources are being explored, but is generally unacceptable for construction purposes when an early indication of risk is required. The chemical test and the gel pat test give results quickly, but the results are not always reliable.

The gel pat test consists of embedding particles of aggregate in neat high-alkali cement paste so that one face of the aggregate particle is exposed at the surface and then immersing the hardened pat in sodium hydroxide solution. If the aggregate is highly reactive exudations of alkali–silicate gel appear on the exposed face of the aggregate within a few days. With aggregates that are of lesser reactivity or that react slowly, gel may not be apparent for some time, if at all.

The chemical test can be made on either fine or coarse aggregate fractions, or, where these are of the same mineralogical composition, on a mixture of both combined in the proportions in which they will be used. The test is made on a sample crushed to between 300 and 150 μm, and consists of reacting the material with a strong, 40 g per litre, solution of sodium hydroxide at an elevated temperature. The solution is analysed to find how much silica was dissolved from the aggregate and by how much the alkalinity of the solution was reduced, that is, how much hydroxide entered into the reaction. The relationship between the two results is used to classify the reactivity of the aggregate.

The test procedures were originally published in 1947.[21] A large number of sands, gravels, rocks and minerals known from the results of mortar bar tests or from their performance in concrete structures to be reactive or non-reactive were tested by the chemical method. It was found that when the amount of silica dissolved was plotted against the reduction in alkalinity, the graph could be divided into areas of known reactivity or non-reactivity. There were, however, some anomalous results, suggesting that the chemical test was not free from interferences. It was found that spurious results for the reduction in alkalinity were given if the aggregate tested contained calcium, magnesium or iron carbonate minerals, such as calcite, dolomite, magnesite or siderite, or contained magnesium silicate minerals, such as serpentine. Chemical tests made on these materials mixed with up to 2.5% of highly reactive opal

Specifications for chemical testing

indicated that the mixtures were innocuous, whereas mortar bars made with the same combinations showed excessive expansions.[22] A third category of reactivity as assessed from the results of the chemical tests was subsequently introduced.[23] This is the troublesome area indicating potentially deleterious aggregates. It is recommended in the ASTM procedure that aggregates so classified by the chemical test should be examined petrographically to determine whether the minerals responsible for spurious chemical test results are present, or whether silica minerals known to be reactive are present. It is also recommended that potential reactivity should be confirmed by measurements of the expansion of mortar bars made with the suspect aggregate diluted over a wide range with an inert aggregate, such as quartz.

When the alkali–silica reaction was first diagnosed as the cause of damage to a number of concrete structures in the USA, a comprehensive programme of tests on UK aggregates was instituted by the Building Research Station.[24] The methods used were petrographic examination, gel pat and chemical tests, and the measurements of the expansion of mortar bars. In the standard method used at that time for mortar bar tests, the alkali content of the cement used in making the specimens was adjusted by the addition of sodium hydroxide solution to the mixing water to give a total equivalent Na_2O content of 1.2% by weight of cement. This has subsequently been modified, and the current ASTM procedure merely stipulates that the cement used should contain more than 0.6% and preferably more than 0.8% of equivalent Na_2O. However, even at the higher content used in the early work, the mortar bars did not show deleterious expansions, although the gel pat and the chemical test did indicate slight reactivity of some aggregates. The overall conclusion from the investigations was that there was little cause for serious concern about alkali–silica reactions occurring with UK aggregates. However, in the past 10 years there has been field evidence that the reaction has occurred in concrete, and a further investigation is being undertaken at the Building Research Establishment to re-examine the test methods currently used and to consider whether there are methods that can be recommended for inclusion in British Standard specifications for concreting aggregates. Preliminary findings have indicated that the

ASTM test methods, particularly the mortar bar test, can give misleading results when applied to UK aggregates, and that expansion tests on concrete specimens could prove more useful.[25]

TEST METHODS FOR CONCRETE

Methods for chemical tests on hardened concrete are given in BS 1881:Part 6.[26] They include the determination of cement content, aggregate/cement ratio, aggregate grading, water/cement ratio and tests for sulphate and chloride contents. Figg and Bowden[27] have given more comprehensive methods of analysis which include the detection or determination of admixtures and contaminants and the examination of chemically attacked concrete. Details are also given of micrometric methods of analysis.

The determination of the mix proportions of hardened concrete is a test frequently requested, but the value of the information to be gained is, in many circumstances, questionable. Certainly, the analysis of concrete that has been substantially altered by chemical action is unlikely to yield accurate information on its original composition. In any circumstances there are many sources of error inherent in the test and these are recognised and identified in the specification. Not least of the errors occurs in sampling. Even when the recommended procedure has been followed, which is that the minimum linear dimension of the sample should be not less than five times the maximum size of the aggregate used, there can be little confidence that a single or a necessarily limited number of samples will be truly representative of the whole mass of concrete placed. Frequently the sample is taken from a point where the concrete looks poor, but that may be due to any of a number of factors concerned with its placing, rather than its composition. Another error, and one which can seriously affect the accuracy of the analysis, arises when the aggregates used are not available for tests to allow corrections to be made for any soluble constituents they may contain. Failure to make the appropriate corrections leads to a calculated cement content greater than the true value. Where essentially acid-insoluble aggregates have been used, for example, flint gravel and quartz sand, the error may not exceed about 10% of the true cement value, but it increases with increasing

Specifications for chemical testing

concentrations of soluble constituents in the aggregates, and can be so great as to make the results meaningless.

The methods given are based on natural aggregates belonging to one of three types, classified according to their solubility in acid. They comprise essentially insoluble materials, such as flint, quartz and quartzite and acid igneous rocks; essentially soluble materials, the limestones and other carbonate rocks; and materials which are partially, and to a varying degree, soluble in acids. The last category, which includes basic igneous rocks – dolerites for example – calcareous sandstones and siliceous limestones, presents the greatest problems in the analysis of concretes in which they have been used. Artificial aggregates are considered as a separate category, and for these, no standard analytical procedures are applicable. It is essential that representative samples of partially soluble aggregates are available before an analysis of the concrete is attempted, but even so, where they have been used in certain combinations, chemical analysis of the concrete is impossible.

The methods, as written, apply to concrete made with Portland cements, but with appropriate modifications to the calculations are applicable to concrete made with other types of cement of standard specified composition. The increasing use of materials such as pulverised fuel ash or ground granulated blastfurnace slag to replace a proportion of Portland cement makes it impossible to determine the mix proportions of concrete incorporating them unless reliable samples of the blended cements that were used are available. This is unlikely to be the case except when the concrete is sampled at an early age.

Two methods for determining mix proportions are given in BS 1881. One is a direct method in which the concrete sample is crushed and ground for the determination of cement content and aggregate/cement ratio. The second also allows for a grading of the aggregate to be made. This requires the coarse aggregate to be separated by hand, an extremely laborious, time consuming, and therefore costly process. The fine aggregate is cleaned of cement by treatment with dilute acid. This results in loss of shell fragments, chalk or other soluble minerals, and can affect the grading of the sand to an extent which may be misleading. The test is obviously inapplicable to concrete made with crushed carbonate rock fine aggregate.

The test for original water content measures the capillary

porosity of the concrete, which is the pore volume originally filled with water at the time of setting of the cement, and the amount of water remaining combined in the cement hydrates. The concrete sample must be sound and neither chemically nor physically damaged. The test is not applicable to poorly compacted concrete nor to concrete which was compacted in a semi-dry condition. The calculation of the original water content requires that the capillary porosity and combined water content of the aggregates used should also be determined. Where samples of the aggregates are not available, it is suggested that coarse aggregate should be separated from another part of the concrete sample and used for the determinations. This approach has its limitations in being only applicable when both coarse and fine fractions are of the same mineralogical type, and in being inapplicable if the coarse aggregate is a conglomerate, for example, a gravel comprising a number of different materials. With very porous aggregates or those that contain appreciable amounts of combined water, the corrections that have to be applied in the calculation of the original water content can give rise to errors so large as to render the results quite unreliable. The accuracy of the determination of the water/cement ratio therefore depends on the accuracy to which the cement content and the original water content of the concrete can be determined and both are dependent on reliable aggregate samples being available for analysis.

The determination of the cement content of concrete can be of use in some circumstances, such as when cubes tested at early ages fail to meet specified strength requirements and it is required to know whether they have been wrongly made or cured or whether the concrete has been incorrectly proportioned. The materials used will then generally be available, and the test results can be viewed with greater confidence than when assumptions have to be introduced into the calculation of cement content. With concrete of greater age, when the original materials are not available, consideration should be given to the question of whether the chemical tests will yield more or better information than can be gained by visual examination of a sawn section of the concrete, by which physical features such as aggregate grading and distribution, and voids due to the use of excessive water or to poor compaction can be recognised.

8
Protection of concrete against chemical attack

Concrete is often required to be placed in situations where it is known that it will be exposed to aggressive conditions and consideration must then be given to its protection. The measures which may be adopted include the use of the more chemically resistant types of cement available or the provision of protective coatings, but the first line of defence is the quality of the concrete. The measures required to achieve good quality lie in the selection of sound materials, in the design of the mix and in the placing and compaction of the concrete. Low permeability is an important factor governing the performance of concrete in any circumstances, and is even more important when the concrete is exposed to chemical attack from external sources. Damaging chemical reactions can take place only in solution, so limiting the penetration of aggressive solutions into the concrete will limit the extent to which attack can occur.

The type of cement used also influences the chemical resistance of the concrete, but although cements have differing degrees of resistance, none is entirely immune to the effects of all aggressive agents at all concentrations, and it is not sufficient to rely solely on their chemical properties to provide protection. A high cement content in the concrete is required, together with a low water/cement ratio. The choice of cements available in the UK has been reduced by the exclusion of high alumina cement from codes of practice and by the unavailability of supersulphated cement, but this has been compensated for by the introduction of pulverised fuel ash and ground granulated blastfurnace slag to be added to Portland cements to give increased chemical resistance. Experience of their performance has been somewhat limited in the UK, although cements based on, or incorporating, these materials have been used extensively in other parts of the world, and in a number of national standards are recommended for use where chemical resistance is required.

The external sources of chemical attack were discussed in Chapter 5. The special circumstances that arise from the use of concrete in tanks or floors in industrial situations are such that no standard approach to protection can be adopted. Lea[1] and Biczok[2] have listed a number of individual chemicals from which concrete is at risk, and various protective methods that have been used. While the materials they recommended may still find a use, modern materials may give better service. A general guide to the protection of concrete by coatings or other chemically resistant materials has been produced by the American Concrete Institute.[3] The subject, however, is one which cannot be covered by general guidelines, and specialist advice should be sought in each case.

ACIDIC SOILS AND WATERS

The protection required when concrete is to be placed in naturally acidic soils and ground waters is a vexing problem and one which has not been entirely resolved. The measurement of soil pH is made on a suspension of soil in water, and this may not truly reflect conditions in situ. The amount of water available in the soil will determine the degree of acidity to which the concrete will be exposed, and if there is little water movement the rate of replenishment of acids from the surrounding soil will be slow, as will be the rate of removal of the reaction products, and hence attack will be limited. Conversely, where water is flowing past the concrete or where the concrete is exposed to a fluctuating water table, the risks are greater. Gutt and Harrison[4] have defined an upper and a lower limit of risk. They consider that a low risk classification would be a pH value of 5.5 or above, a stiff unfissured clay soil and a water table below foundation level, and that in such conditions no particular precautions would be necessary. Their high risk classification would be where the pH value is less than about 3.5, where there is some movement of water adjacent to the concrete and a high water table creating a hydrostatic head. These conditions would require protective measures, such as coatings or linings to the concrete. Between these extremes there remains the problematical area where the risks are less readily assessed and where there is little published guidance on precautions that may be required. Measures that have been proposed or adopted are to increase the cement content of the concrete, and thereby its impermeability, to increase the dimensions of the concrete, providing a sacrificial layer,[5] or to pack crushed limestone around the exposed face of the concrete to neutralise the acid.[6] The last two measures raise the question of how much to allow, and although the question may be answered theoretically from an accurate knowledge of the water table and water movement, observations of these are usually made over a limited period of time, and do not reflect seasonal variations in ground water levels, nor can they predict future water movements, for example, along loosely backfilled trenches.

Blastfurnace slag and pozzolanic cements are recommended in European standard specifications for concrete to be placed in acidic conditions. Their increased resistance has been attributed to their lower contents of calcium hydroxide, which is the constituent of Portland cement most vulnerable to acid attack and to leaching.

A recent French standard[7] allows the use of Portland cement where pH values are between 6.5 and 4.5, provided that the cement has low contents of tricalcium silicate and tricalcium aluminate, although the standard does not specify how low. Since tricalcium silicate releases more calcium hydroxide into solution than does dicalcium silicate, the stipulation reflects the view that increased resistance is given when the content of calcium hydroxide in the cement is reduced. The requirement for a low tricalcium aluminate content comes about because sulphate contents are included in the assessment of aggressiveness, together with the pH value and the content of aggressive carbon dioxide.

No cementitious materials can withstand acid conditions represented by low pH values of less than about 4.0, and concrete must then be given positive protection by the use of acid-proof membranes or coatings. Admixtures are of little value, although the use of water-reducing admixtures results in increased impermeability of the concrete, provided that a high cement content is maintained. Integral waterproofers are unlikely to provide sufficient protection. However, the use of polymers in the mix is allowed in the French standard where concrete is to be placed in ground waters classified as extremely aggressive. This is as an alternative to providing the concrete with an impermeable, protective coating. Waters that are considered to be extremely aggressive with regard to their acidity are those with a pH value of less than 4.0, or an aggressive carbon dioxide content greater than 100 mg/litre.

SULPHATES

Recommendations for concrete to be exposed to the action of sulphates in soils and ground waters are given in BRE Digest 250[8] and in the more recently published BS 8110,[9] both of which classify aggressiveness in the same five categories of increasing severity. There are slight differences between the two sets of recommendations. In BS 8110, Class 1 sulphate conditions are considered to be non-aggressive, and no general requirements are given for the cement content or water/cement ratio of concrete. It is stated that for foundations to low rise structures, the concrete can be either of a minimum strength grade of C20, provided that the cement content is not less than 220 kg/m^3, or of strength grade C10P, both of which are defined in BS 5328.[10] In Digest 250, cement contents and water/cement ratios are given for plain and for reinforced concrete to be placed in Class 1 conditions, although there is a rider that a relaxation of both is permissible for concrete in strip foundations and in trench fill. More types of cement are listed in BS 8110, including Portland pulverised fuel ash cement[11] and supersulphated cement.[12]

Both documents recommend the addition of pulverised fuel ash or ground granulated blastfurnace slag to ordinary Portland cement for increased resistance against moderate concentrations of sulphates. The resistance depends on a fairly large replacement of cement – more than 70% is required in the case of blastfurnace slag and more than 25% if pulverised fuel ash is used. The use of blastfurnace slag with sulphate-resisting Portland cement results in improved resistance to moderate sulphate concentrations, but where conditions are more severe, sulphate-resisting Portland cement alone is recommended. It is stated in BS 8110 that pulverised fuel ash should not be used in combination with sulphate-resisting Portland cement in concrete required to be resistant to the action of sulphates. A summary of data given in BS 8110 is shown in Table 9.

CHLORIDES

The corrosion of reinforcing steel is a major cause of damage to concrete and can occur whenever air and water are able to penetrate through the cover and diffuse to the steel. Under nor-

Table 9. *Requirements for concrete exposed to the action of sulphates*

Degree of severity of exposure	Sulphate content (as SO_3)			Type of cement recommended*	Cement content, kg/m³
	Soils				
	Total SO_3, %	SO_3 in 1:2 soil: water extract, g/l	Water, g/l		
Moderate	0.2–0.5	1.0–1.9	0.3–1.2	Any, plus OPC with PFA or GGBFS	330
				OPC + 25–40% PFA OPC + 70–90% GGBFS	310
				SRPC or SSC	280
Severe	0.5–1.0	1.9–3.1	1.2–2.5	OPC + 25–40% PFA OPC + 70–90% GGBFS	380
				SRPC or SSC	330
Very severe	1.0–2.0	3.1–5.6	2.5–5.0	SRPC or SSC	370
Extreme	>2.0	>5.6	>5.0	SRPC or SSC with protective coatings or linings	370

* OPC = ordinary Portland cement (BS 12), SRPC = sulphate-resisting Portland cement (BS 4027), SSC = supersulphated cement (BS 4248), PFA = pulverised fuel ash (BS 3892), GGBFS = ground granulated blastfurnace slag. Including in the types of cement that may be used in conditions of moderate severity are Portland–blastfurnace cements to BS 146 and BS 4246 and Portland pulverised fuel ash cement to BS 6588.

mal conditions, the steel is protected both by the alkalinity of the cement paste and by the provision of an adequate depth around it of concrete of low permeability. However, in the presence of chlorides, whether these are introduced into the concrete with the materials used or enter from external sources, the risk of corrosion is greatly increased, as discussed in Chapter 5. Where limits set[13] for contents of chloride in the concrete cannot be met because the materials specified are not available, or where exposure conditions are severe, direct methods to protect the reinforcement may need to be considered to prevent damage to the concrete. These include the use of corrosion resistant metals and the application of coatings to the reinforcement or to the structure.

The cost of inherently corrosion-resistant metals, such as stainless steel alloys, is likely to limit their use to situations where it is not possible to provide a normal depth of cover, such as in thin domes or cladding panels. Metallic coatings, generally of zinc, have been used with satisfactory results to provide sacrificial protection to steel reinforcement under normal conditions of atmospheric corrosion, but, in the presence of chlorides, the zinc suffers increased attack. Although reports on the ability of galvanised zinc coatings to protect steel against chloride corrosion are conflicting, the consensus of opinion is that if conditions are sufficiently severe for the steel to be corroded, the zinc will rapidly be consumed sacrificially, and deterioration of the concrete will only be delayed; it will not be prevented. This view is supported by work at Building Research Establishment[14] where it was found that when concrete contained high levels of chloride, 2% or more, cracking was not significantly different whether galvanised or bare steel reinforcement were used. It was considered that the small delay before the onset of cracking could make the use of galvanised steel not economically worth while.

Organic coatings have also been used for the protection of reinforcement. These range from simply applied bitumen coats or paint systems, neither of which is entirely satisfactory since their use can result in a loss of bond between the steel and the concrete, to factory-applied coatings. Powder epoxy coated reinforcement has been used widely in recent years in the USA in concrete bridge decks where there is a risk of corrosion from the use of chloride de-icing salts. Observations have indicated a significant reduction in the incidence of cracking of the cover. Other specially formulated organic coatings have also been proposed. One proprietary system, designed to protect reinforcing bars during transit from the mill to the construction site, consists of coating the steel with an alkali-resistant polyvinylbutyral etch primer. It is reported that this coating is flexible, allows cold bending and has good bond characteristics.[15]

A method of protecting steel which has had considerable application is to coat the bars on site with a slurry of cement and water, to give a layer of iron and calcium-rich compounds on the metal surface. A polymer-modified coating of cement, silica fume and fine sand has been reported to offer effective protection.[16] It is claimed that this provides a dense, continuous coat of low porosity around the steel and has a good bond to both the steel and the concrete. Organic corrosion inhibitors are included in the formulation and since these are applied around the steel, better continuity of contact with metal might be expected than when inhibitors are mixed into the concrete, as discussed below.

The protection provided by any coating is dependent on its continuity and its uniform adhesion, which are in large measure governed by the condition of the surface to which it is applied. Where imperfections occur during application or where damage has resulted from bending or handling on site, corrosion of the steel will be concentrated at small anodic areas and consequently will be deeper. An advantage of epoxy coatings is that any damage sustained before placing can be made good with a suitable epoxy paint.

The protection of reinforcement from chloride corrosion by coating the concrete, rather than the steel, is a complex and much debated subject, of which only brief mention will be made here. A primary consideration is the extent to which it is practicable, or possible, to apply a uniform coating to all exposed surfaces of a structure. Any gaps will have the effect of promoting localised pitting corrosion in the same way as do imperfections in coatings applied to the reinforcement. The condition of the concrete when the coating is applied is also a factor in determining the success of the measure. If much water is sealed within concrete containing chlorides, corrosion may still proceed from the operation of differential chloride concentration cells, even after oxygen originally present has been exhausted. Coatings are unlikely to be successful, either in application or performance, when they are used on poor quality concrete.

Chemicals that inhibit the corrosion of metals are widely used to provide protection to steel in water-holding and heating systems and it might seem appropriate for them to be used in concrete to prevent the corrosion of the reinforcing steel. In practice, however, they have not proved entirely successful. One problem is to ensure that the inhibitor forms a continuous film over the entire surface of the metal. If the film is not continuous, the protection given will be only partial and local, and corrosion will be concentrated at the less inhibited areas, thereby promoting pitting of the steel. The difficulties of ensuring that the steel is uniformly protected are increased when chlorides are present in the concrete. Whereas quite small concentrations of inhibitors are adequate to protect steel vessels holding fresh water, very much higher amounts are required to be effective if the water contains even low contents of sodium or other chlorides. The distribution of chlorides is generally not uniform throughout concrete, and the problem then arises of ensuring that the amount of inhibitor used is sufficient to overcome the corrosive action on the steel of pockets of high chloride concentrations. Inhibitors might also have the effect of loosening any scale on the reinforcing bars and so weakening the bond to the concrete. Moreover, in the high concentrations required, some corrosion-inhibiting chemicals, notably sodium nitrite and sodium benzoate, have the adverse effect of reducing the compressive strength of the concrete.

SEA WATER
Concrete placed in sea water is at risk of chemical attack, of damage resulting from the crystallisation of salts within it under conditions of alternate wetting and drying, of corrosion of any reinforcing steel and of mechanical damage due to wave action. Attack by frost is a serious cause of failure in very cold waters. When it has been damaged in any of these ways, the concrete is more susceptible to attack by the other destructive agents.

The chemical reactions that occur between Portland cement and magnesium sulphate present in the sea water were discussed in Chapter 5. An equally important factor in deterioration is corrosion of the steel in reinforced concrete. This is most likely to occur in the inter-tidal zone, above high water level or where the concrete is exposed to sea spray, as on the underside of decks. Concrete in these water level zones is also at risk from the effects of salts carried upwards by capillary action. The higher concentrations that result increase the rate of chemical attack on the cement paste, and as the water dries out, the concrete may suffer physical disruption due to crystallisation of the salts. Corrosion is less likely where the concrete is permanently submerged, since the diffusion of oxygen is slow through saturated concrete, and since conditions surrounding the steel are generally uniform.

Despite the risks, concrete has been used extensively and successfully in maritime works. Pozzolanic cements are frequently used in Mediterranean countries and blastfurnace slag cements are used in northern European countries. Ordinary Portland cement concretes have given satisfactory service in temperate waters, particularly where they are fully submerged, and it has been suggested[4] that the increased resistance to chemical attack that may be gained by the use of sulphate-resisting Portland cement is probably not required in these conditions if the quality of the concrete can be assured. Coatings based on tar or bituminous materials have been tried as protective measures for concrete placed in sea water, but their effective life has been found to be too short for them to be of practical use, particularly above low water level where they are exposed to harsh erosive conditions. They have, however, given acceptable protection to concrete exposed only to the action of spray.

ALKALI-SILICA REACTION

The deterioration of concrete as the result of chemical reactions between certain types of siliceous mineral in aggregates and alkali metal hydroxides was discussed in Chapter 3. It is not possible to remove reactive material from an aggregate by any form of processing, but when a particular type is known to have caused damage or when tests have indicated possible reactivity, certain precautionary measures can be adopted to reduce the risk of damage to the concrete. Alkali and water are needed for expansion to take place, so limiting either of these will reduce the risk. The chief source of alkali is Portland cements, although there is some evidence that alkali metal salts from external sources, for example sea water or de-icing salts, may contribute to the reaction. A limit of 0.6% equivalent Na_2O of alkali in the cement has been found by experience to be the level below which expansion is unlikely to occur, and although British Standard specifications for cements do not include limits for alkali content, in some other countries where volcanic rocks form a large part of the natural aggregates, the limit of 0.6% equivalent Na_2O is set when the cement is to be used in combination with materials known to be reactive.

The alkali contents of ordinary Portland cements manufactured in the UK range from below 0.6 to above 1.0% equivalent Na_2O, but levels consistently below 0.6% are not guaranteed by the producers. Sulphate-resisting Portland cement having an equivalent Na_2O content guaranteed not to exceed 0.6% can be supplied, provided that a low alkali content is stipulated at the time when the cement is ordered. This option is also available for ASTM cements types II and V. Although a safe level for the alkali content of the cement has been established, a value based on the alkali content of the concrete is considered to be more realistic, since it takes into account the cement content of the mix, and an equivalent Na_2O content in the concrete of 3 kg/m^3 has been found to be the limit below which damaging expansion is unlikely to occur.[17,18]

The alkali content of the concrete can be reduced by the use of blastfurnace slag cements or cement replacement materials such as pulverised fuel ash or ground granulated blastfurnace slag. These contain alkali metals in concentrations which may exceed those of ordinary Portland cements, and it might therefore seem that their use to replace part of the cement would not have a significant effect in reducing the alkali content of the concrete. It appears, however, that the alkali metals are less readily released from the replacement materials than from Portland cement, and their use is recommended in BRE Digest 258[19] and in the report of a working party on the problem of alkali–silica reactions.[20] However, the effectiveness of pulverised fuel ash in all cases has been questioned, and Palmer[21] has reported that damaging expansion to concrete can sometimes occur when amounts of up to 40% are used to replace Portland cement. Opinions are also divided on the value of natural pozzolanas. Stanton[22] found some to be effective, but also found that some, while inhibiting expansion with one type of aggregate, could accelerate the expansive effect of another. Sims and Poole[23] similarly found that mortar bars made with an aggregate known to be reactive expanded significantly when pozzolana was used to replace part of the cement. In this case, only 10% of pozzolana was used, which might have been insufficient to be effective. Stanton considered that as much as 40% was necessary. Natural pozzolanas are included in BRE Digest 258 as cement replacement materials, although the amounts to be used are not given. It is made plain in Digest 258 that the recommendations for cement replacements refer only to use with UK aggregates, and that for overseas work the effect of any replacements should be assessed by comparing the length changes of concrete or mortar specimens made with the cement and the aggregates proposed for use, with and without the replacement materials.

Water is necessary for any silicate gel formed to be able to swell to disruptive volume, and any measures that prevent ingress of water into the concrete are beneficial. In most cases of damage, the concrete has been placed in situations where water can move through it, as in dams, ground slabs and foundations with their surfaces exposed to the atmosphere, or in structures partially immersed in water, such as bridge and jetty piers. Sufficient water for the reaction can be provided in exposure conditions of heavy condensation, high humidity or frequent wetting. There is evidence that where water can migrate through concrete, alkalis in solution can become concentrated close to the surface from which the water evaporates, and localised attack can occur to a greater extent than might be expected from the alkali content of the concrete as a whole.[24]

SURFACE PROTECTION OF CONCRETE

In some conditions of exposure it is necessary to provide concrete with external forms of protection against chemical attack by aggressive agents. Most generally these are required when the concrete is to be placed in soils or waters containing high concentrations of damaging substances, but other circumstances might arise when the exclusion of water or of air or other gases in the atmosphere can give increased durability to a structure. Examples, which have been mentioned earlier, are to reduce water penetration into concrete made with aggregates that are likely to enter into alkali-silica reactions, and to protect reinforcing steel against corrosion in concrete that contains high levels of chlorides. The protection given may take the form of either linings or coatings. The practical problems of their successful application to all exposed surfaces are not discussed here, where only the chemical properties of the materials that may be used are considered.

Except in industrial situations, and in pipes, drains or culverts carrying aggressive solutions, the use of linings is generally confined to the protection of concrete in foundations. Sheets of flexible materials, such as polyethylene, unplasticised polyvinyl chloride or synthetic rubbers are mainly used. These may be jointed by heat welds, glues or folded overlaps. These materials are unaffected by sulphates, chlorides or naturally occurring acids, but are not always resistant to the effect of certain organic chemicals that may be found during the redevelopment of contaminated land, notably the sites of former gas works. The protection of concrete in piled foundations is a special case of the use of permanent liners, which may be either flexible or rigid. Bartholomew[25] has described a number of measures that have been adopted for use in soils containing high concentrations of sulphates, chlorides and acids.

Coatings are used both below and above ground level. The choice of materials available is very wide and they can only be treated here in terms of their general composition and application. The selection of any particular coating will depend on the degree of protection dictated by the conditions to which the concrete will be exposed and the appearance required of the structure. The basic requirements of any coating are that it should have good adhesion to the concrete, which is often wet when the treatment is applied, that it should be resistant to penetration by aggressive solutions, be free of pinholes and not liable to crack in service, be easy to apply under site conditions, and, when used below ground, be resistant to microbial degradation. Below ground level, protective coatings are mainly bitumen based. In severe conditions, such as where water pressures are involved, thick, hot-trowelled coats of mastic asphalt are often applied. Bituminous paints, in which the bitumen is dissolved in a solvent, and emulsions of bitumen in water, give only thin surface coatings, which have limited resistance to abrasion and weathering. Membranes of glass fibre fabric are frequently used as reinforcement to bituminous coatings. When permanently below water, or in waterlogged soils, bituminous coatings are apt to swell and spall. They may also be subject to damage caused by microbial activities.

A wide range of protective materials is available for use on superstructure concrete, ranging from water repellent treatments to thick paint systems. The water repellent materials are generally colourless, and consist either of solutions of hydrophobic substances which act by lining the pores of the concrete, or of solutions of chemicals which react with constituents of the cement to form insoluble compounds. Silicones and oils act in the former manner. Examples of materials which react on the concrete surface are solutions of alkali metal soaps, which reduce permeability by the formation of insoluble calcium soaps with the calcium hydroxide present in the cement, and silicofluorides, generally of zinc or magnesium, which form insoluble calcium fluorides and also act on the silicates of the cement to produce insoluble forms of calcium silicate.

Paints based on drying oils, such as linseed oil, cannot be successfully applied to fresh concrete. The oils are saponified by alkalis present in the cement and the paint remains soft and sticky. The film may also blister owing to water pressure from within the concrete. However, paints in which the medium is non-saponifiable and which can therefore be used on alkaline surfaces are available. These include chlorinated rubber paints, paints based on synthetic resins, for example, acrylic polymers, and various co-polymers, of which styrene-butadiene has proved to have good adhesion to concrete. These paints can be produced to give thick-textured finishes by the inclusion of sand, mica or other mineral fillers in the formulatin.

The term cement paint refers to paint based on Portland cement and containing fillers and pigments. Water repellent substances, such as calcium or aluminium stearate, are often included to improve the waterproofing properties of cement paints, and set accelerators, generally calcium chloride, are added to ensure that the cement hydrates before the paint dries out. Improved flexibility of the coating may be gained by the inclusion of polymer resins. Cement paints are widely used on external surfaces for decorative purposes and to prevent penetration by rain.

Appendix 1. Atomic and molecular weights and molecular volumes

Element	Symbol	Atomic weight
Aluminium	Al	26.98
Calcium	Ca	40.08
Carbon	C	12.01
Chlorine	Cl	35.457
Hydrogen	H	1.008
Iron	Fe	55.85
Magnesium	Mg	24.32
Manganese	Mn	54.94
Oxygen	O	16.00
Potassium	K	39.10
Silicon	Si	28.09
Sodium	Na	22.99
Sulphur	S	32.06

Compound	Formula	Molecular weight
Aluminium oxide	Al_2O_3	101.96
Calcium aluminate	$CaO \cdot Al_2O_3$	158.04
Calcium carbonate	$CaCO_3$	100.09
Calcium hydroxide	$Ca(OH)_2$	74.10
Calcium sulphate	$CaSO_4$	136.14
Calcium sulphate (gypsum)	$CaSO_4 \cdot 2H_2O$	172.18
Carbon dioxide	CO_2	44.01
Dicalcium silicate	$2CaO \cdot SiO_2$	172.25
Iron oxide (ferric)	Fe_2O_3	159.70
Silica	SiO_2	60.09
Sulphuric anhydride (sulphate)	SO_3	80.06
Tetracalcium aluminoferrite	$4CaO \cdot Al_2O_3 \cdot Fe_2O_3$	485.98
Tricalcium aluminate	$3CaO \cdot Al_2O_3$	270.20
Tricalcium silicate	$3CaO \cdot SiO_2$	228.33

Compound	Molecular volume, cm^3
Calcium hydroxide	33
Calcium sulphate (gypsum)	74
Tricalcium aluminate	150
Tricalcium monosulphate sulphoaluminate	313
Tricalcium trisulphate sulphoaluminate (ettringite)	715

Appendix 2. Abbreviated formulae used in cement chemistry

C	=	CaO, calcium oxide
A	=	Al_2O_3, aluminium oxide
S	=	SiO_2, silica
F	=	Fe_2O_3, iron oxide (ferric)
K	=	K_2O, potassium oxide
N	=	Na_2O, sodium oxide
H	=	H_2O, water

hence

C_3S	=	$3CaO \cdot SiO_2$, tricalcium silicate
C_2S	=	$2CaO \cdot SiO_2$, dicalcium silicate
C_3A	=	$3CaO \cdot Al_2O_3$, tricalcium aluminate
C_4AF	=	$4CaO \cdot Al_2O_3 \cdot Fe_2O_3$, tetracalcium aluminoferrite

AFm phases are phases typified by the low (mono) sulphate form of tricalcium sulphoaluminate, $C_3A \cdot CaSO_4 \cdot 12H_2O$, in which the Al can be substituted by Fe and the SO_4 by other anions, e.g. Cl.

AFt phases are typified by the high (tri) sulphate form of tricalcium sulphoaluminate (ettringite), $C_3A \cdot 3CaSO_4 \cdot 32H_2O$, in which the Al can be substituted by Fe and the SO_4 by other anions.

References

CHAPTER 1
1. British Standards Institution. *Methods for specifying concrete including ready-mixed*. BSI, London, 1981, BS 5328.
2. British Standards Institution. *Structural use of concrete. Part 1:Code of practice for design and construction*. BSI, London, 1985, BS 8110.
3. Department of the Environment. *Design of normal concrete mixes*. Her Majesty's Stationery Office, London, 1975.

4. Murdock L.J. and Brook K.M. *Concrete materials and practice*, 5th edn. Edward Arnold, London, 1979.
5. Neville A.M. *Properties of concrete*, 3rd edn. Pitman, London, 1981.
6. Orchard D.F. *Concrete technology*, Vols 1–3, 4th edn. Applied Science, London, 1979.
7. British Standards Institution. *Methods for sampling and testing of mineral aggregates, sands and fillers*. BSI, London, 1983, BS 812.
8. British Standards Institution. *Specification for aggregates from natural sources for concrete*. BSI, London, 1983, BS 882.
9. Collis L. and Fox R.A. (eds) *Aggregates: sand, gravel and crushed rock aggregates for construction purposes*. The Geological Society, London, 1985, Geological Society Engineering Geology Special Publication No. 1.

CHAPTER 2

1. Lea F.M. *The chemistry of cement and concrete*, 3rd edn. Edward Arnold, London, 1970, 8.
2. Taylor H. Modern chemistry of cements. *Chem. Ind.*, 1981, 19 September, 620–625.
3. Diamond S. Cement paste microstructure – an overview at several levels. *Proc. conf. on hydraulic cement pastes: their structures and properties, University of Sheffield, April 1976.* Cement and Concrete Association, Wexham Springs, Slough, UK, 1976, 2–30.
4. Building Research Establishment. *Materials for concrete.* Her Majesty's Stationery Office, London, 1980, Digest 237.
5. British Standards Institution. *Sulphate-resisting Portland cement.* BSI, London, 1980, BS 4027.
6. American Society for Testing and Materials. *Standard specification for Portland cement.* ASTM, Philadelphia, 1981, C150–81.
7. British Standards Institution. *Ordinary and rapid hardening Portland cement.* BSI, London, 1978, BS 12.
8. British Standards Institution. *Portland–blastfurnace cement.* BSI, London, 1973, BS 146.
9. British Standards Institution. *Low heat Portland–blastfurnace cement.* BSI, London, 1974, BS 4246.
10. British Standards Institution. *Supersulphated cement.* BSI, London, 1974, BS 4248.
11. British Standards Institution. *Structural use of concrete. Part 1:Code of practice for design and construction.* BSI, London, 1985, BS 8110.
12. British Standards Institution. International Standards Organisa-

References

tion ISO/R597, given in BS 4627. *Glossary of terms relating to types of cements and their properties.* BSI, London, 1970.
13. British Standards Institution. *Part 2:Methods of testing cements. Chemical tests.* BSI, London, 1970, BS 4550.
14. British Standards Institution. *Pulverized-fuel ash. Part 1:Specification for pulverized-fuel ash for use as a cementitious component in structural concrete.* BSI, London, 1982, BS 3892.
15. British Standards Institution. *Portland pulverized-fuel ash cement.* BSI, London, 1985, BS 6588.
16. British Standards Institution. *Pozzolanic cement with pulverized-fuel ash as pozzolana.* BSI, London, 1985, BS 6110.
17. American Society for Testing and Materials. *Fly ash and raw or calcined natural pozzolana for use as a mineral admixture in Portland cement concrete.* ASTM, Philadelphia, 1980, C618–80.
18. Rio A. and Celani A. Levobuzioue dei leganti pozzolanici alla luce di recenti applicazioni in opare esposte all 'acqrea di mare. *International symposium on the behaviour of concretes exposed to sea water, Palermo, Sicily, 1965*, Réunion Internationale des Laboratoires d'Essais et de Recherches sur les Matériaux et les Constructions, Paris, 1985.
19. British Standards Institution. *High alumina cement.* BSI, London, 1972, BS 915.
20. Lea F.M. *The chemistry of cement and concrete*, 3rd edn. Edward Arnold, London, 1970, 507.
21. Lea F.M. *Some studies on the performance of concrete in sulphate-bearing environments.* Canadian Building Series. Performance of Concrete (ed. E.G. Swenson), University of Toronto Press, 1968.
22. Lafuma H. Calcium aluminates in aluminous and Portland cements. Lecture to the Istituto Eduardo Torroje, Madrid, May 1963.

CHAPTER 3
1. Building Research Establishment. *Shrinkage of natural aggregates in concrete.* Her Majesty's Stationery Office, London, 1968, Digest 35.
2. Cement and Concrete Association. *Impurities in aggregates for concrete.* C&CA, Wexham Springs, Slough, 1970, Advisory Note 18.
3. American Society for Testing and Materials. *Standard specification for concrete aggregates.* ASTM, Philadelphia, 1981, Designation C33–81.

4. British Standards Institution. *Methods for specifying concrete including ready-mixed.* BSI, London, 1981, BS 5328.
5. Fookes P.G. and Collis L. Cracking and the Middle East. *Concrete*, 1976, Feb. 14–19.
6. Samarai M.A. The disintegration of concrete containing sulphate-contaminated aggregates. *Mag. Concr. Res.*, 1976, **28**, Sept. 130–147.
7. Wimpey Laboratories Limited. *Technical Report CC32.* Wimpey, Hayes, 1985.
8. Whitely P., Russman H.D. and Bishop J.D. Porosity of building materials – a collection of published results. *J. Oil and Colour Chemists Assoc.* 1977, **60**, 142–150.
9. Poole A.B. and Thomas A. A staining technique for the identification of sulphates in aggregates and concrete. *Mineral Mag.*, 1975, **40**, Sept. 315–316.
10. British Standards Institution. *Methods for sampling and testing of mineral aggregates, sands and fillers.* BSI, London, 1976, BS 812:Amendment 4617, 1984, *Methods for detection of chemical properties.*
11. Rosa E.B., McCollom B. and Peters O.S. *Electrolysis in concrete.* United States Bureau of Standards Technical Paper 18, 1913.
12. Building Research Establishment. *The durability of steel in concrete. Part 1:Mechanism of protection and corrosion.* Her Majesty's Stationery Office, London, 1982, Digest 263.
13. Roberts M.H. The effect of calcium chloride on the durability of pre-tensioned wire in prestressed concrete. *Mag. Concr. Res.*, 1962, **14**, Nov., 143–154.
14. Greater London Council. *Marine aggregates specification.* GLC, London, 1968, Bulletin 16.
15. British Standards Institution. *Structural use of concrete. Part 1:Code of practice for design and construction.* BSI, London, 1985, BS 8110.
16. Figg J.W. and Lees T.P. Field testing the chloride content of sea-dredged aggregates. *Concrete*, 1975, **9**, Sept., 38–40.
17. Stanton T.E. The expansion of concrete through reaction between cement and aggregate. *Proc. Am. Soc. Civ. Engrs*, 1940, **66**, Dec., 1781–1811.
18. Jones F.E. and Tarleton R.D. *Reactions between aggregates and cement.* Her Majesty's Stationery Office, London, National Building Studies Research Papers 14, 15, 17 (1952) and 20, 25 (1958).
19. Gutt W. and Nixon P. Alkali–aggregate reaction in concrete in the UK. *Concrete*, 1979, **13**, May, 19–31.

References

20. Palmer D. Alkali–aggregate reactions in Great Britain – the present positions. *Concrete*, 1981, **15**, March, 24–27.
21. Vivian H.E. Alkali–aggregate reaction. *Symposium on alkali–aggregate reaction – preventative measures, Reykjavik, August 1975*. Rannsobknastofnun Byggingandnadar, 1985.
22. Building Research Establishment. *Alkali aggregate reactions in concrete*. Her Majesty's Stationery Office, London, 1982, Digest 258.
23. Jones F.E. Alkali aggregate interaction in concrete. *Chem. Ind.*, 1953, 26 Dec. 1375–1382.
24. National Research Council of Canada. *Alkali aggregate reactions in Nova Scotia*, Parts I–IV. NRCC, Ottawa, 1973.
25. Swenson E.G. A reactive aggregate undetected by ASTM tests. *ASTM Bull.* 1957, No. 226, Dec. 48–51.
26. Gillot J.E. and Swenson E.G. Mechanism of the alkali carbonate rock reaction. *Quart. J. Engng Geol.*, 1969, **2**, Oct., 7–20.
27. Midgley H.G. The staining of concrete by pyrite. *Mag. Concr. Res.*, 1958, **10**, Aug., 75–78.
28. Moum J. and Rosenqvist I Th. Sulphate attack on concrete in the Oslo region. *J. Am. Concr. Inst.*, 1959, Sept., 257–264.
29. Fristrom G. and Sallstrom S. Control and maintenance of concrete structures in existing dams in Sweden. *Proc. 9th international conference on large dams, Istanbul, 1967*, **3**, 383–410.
30. Hageman T. and Roosnan H. Chemical reactions involving aggregate. Paper VI–I, *Fourth international symposium on the chemistry of cement, Washington DC, 1960*. US Department of Commerce, National Bureau of Standards, Monograph 43, Vol. 2.
31. Ruxton B.P. and Berry L. Weathering of granite and associated erosional features in Hong Kong. *Bull. Geol. Soc. America*, 1967, **68**, Oct., 1263–1292.
32. Lea F.M. *The chemistry of cement and concrete*, 3rd edn. Edward Arnold, London, 1970, 568.
33. Orchard D.F. Concrete technology, Vol. 1:Properties of materials, 4th edn. Applied Science, London, 1979, 144.
34. Schnitzer M. and Khan S.U. *Humic substances in the environment*. Dekker, New York, 1972.
35. Paul E.A., Campbell C.A., Rennie D.A. and McCallum K.J. *Eighth international congress of soil science, Budapest, 1964*, 201–208.
36. British Standards Institution. *Specification for aggregates from natural sources for concrete*. BSI, London, 1983, BS 882.
37. American Society for Testing and Materials. *Organic impurities in sands for concrete*. ASTM, Philadelphia, 1973, C40–73.

38. American Society for Testing and Materials. *Effects of organic impurities in fine aggregates on the strength of mortar.* ASTM, Philadelphia, 1975, C87–69.
39. American Society for Testing and Materials. *Standard specification for concrete aggregates.* ASTM, Philadelphia, 1981, C33–81.
40. Schnitzer M. *Proc. international symposium on soil organic matter.* 3rd Braunschweig International Atomic Energy Agency, Vienna, 1977, 117–132.
41. Pairon G. *Cimenteries CBR Cementbedrijven*, 1976. Personal communication.
42. Nurse, R.W. and Midgley H.G. The mineralogy of blastfurnace slag. *Silic. Ind.* 1951, **16**, No. 7, 211–217.
43. British Standards Institution. *Air-cooled blastfurnace slag aggregate for use in construction.* BSI, London, 1983, BS 1047.
44. Everett L.H. and Gutt W. Steel in concrete with blastfurnace slag aggregate. *Mag. Concr. Res.*, 1967, **19**, June, 83–94.
45. British Standards Institution. *Part 2:Foamed or expanded blastfurnace slag lightweight aggregate for concrete.* BSI, London, 1973, BS 877.
46. British Standards Institution. *Part 2:Methods for sampling and testing of lightweight aggregates.* BSI, London, 1973, BS 3681.
47. British Standards Institution. *Clinker and furnace bottom ash aggregate for concrete.* BSI, London, 1985, BS 1165.

CHAPTER 4
1. British Standards Institution. *Methods of test for water for making concrete (including notes on the suitability of the water).* BSI, London, 1980, BS 3148.
2. British Standards Institution. *Structural use of concrete. Part 1:Code of practice for design and construction.* BSI, London, 1985, BS 8110.
3. Grieb W.E., Werner G. and Woolf D.O. Water-reducing retarders for concrete – physical tests. *Public Roads*, 1961, **31**, Feb., 136–154.
4. British Standards Institution. *Concrete admixtures. Part 1:1982. Specification for accelerating admixtures, retarding admixtures and water-reducing admixtures. Part 2:1982. Specification for air-entraining admixtures. Part 3:1985. Specification for superplasticizing and retarding superplasticizing admixtures.* BSI, London, BS 5075.
5. *Admixture data sheet.* Cement Admixtures Association Ltd, London, 1975.
6. Rixom M.R *Chemical admixtures for concrete.* Wiley, New York, 1978.
7. Orchard D.F. *Concrete technology, Vol. 1. Properties of materials*, 4th edn. Applied Science, London, 1979.

References

8. Ramachandran V.S. *Calcium chloride in concrete. Science and technology.* Applied Science, London, 1976.
9. Shideler T.J. Calcium chloride in concrete. *Proc. Am. Concr. Inst.*, 1952, **48**, 537–560.
10. Lea F.M. *The chemistry of cement and concrete*, 3rd edn. Edward Arnold, London, 1970, 605.
11. Illston J.M. Aspects of the behaviour of the cement paste phase of composite materials, with reference to practical problems of concrete technology. *Proc. Conf. on hydraulic cement pastes: their structure and properties, University of Sheffield, April 1976.* Cement and Concrete Association, Wexham Springs, Slough, 1976, 239.

CHAPTER 5
1. Building Research Establishment. *Concrete in sulphate-bearing soils and ground waters.* Her Majesty's Stationery Office, London, 1981, Digest 250.
2. Lea F.M. *The chemistry of cement and concrete*, 3rd edn. Edward Arnold, London, 1970, 345.
3. British Standards Institution. *Structural use of concrete. Part 1:Code of practice for design and construction.* BSI, London, 1985, BS 8110.
4. Building Research Establishment. *Hardcore.* Her Majesty's Stationery Office, London, 1983, Digest 276.
5. Pourbaix M. Applications of electrochemistry in corrosion science and practice. *Corros. Sci.*, 1974, **14**, Jan. 25–82.
6. Gutt W. and Nixon P. Alkali–aggregate reaction in concrete in the UK. *Concrete*, 1979, **13**, May, 19–21.
7. Palmer D. Alkali–aggregate reaction in Great Britain – the present position. *Concrete*, 1981, **15**, Mar., 24–27.
8. Osborne G.J. The durability of lightweight concrete made with pelletized slag as aggregate. *Durability of building materials*, 1985, **2**, 249–263.
9. Halstead P.E. An investigation of the erosive effect on concrete of soft water of low pH value. *Mag. Concr. Res.*, 1954, **6**, Sept., 93–98.
10. Kovalenko N.P., Evseev G.A., Ploskonsov V.N. and Antonov V.V. Corrosion of concrete in corrosive peat ground waters. *Stroit Architekt*, 1971, **14**, No. 3, 85–87.
11. Eglinton M.S. *Review of concrete behaviour in acidic soils and ground waters.* Construction Industry Research and Information Association, London, 1975, CIRIA Technical Note 69.
12. Penner E., Eden W.J. and Grattan–Bellew P.E. *Expansion of pyritic shales.* National Research Council of Canada, Ottawa,

Canadian Building Digest CBD 152.
13. Lea F.M. Some studies on the performance of concrete structures in sulphate-bearing environments. In *Performance of concrete.* University of Toronto Press, 1968, 56–65.
14. Penner E., Eden W.J. and Gillott J.E. Floor heave due to biochemical weathering of shale. *Proc. 8th international conference on soil mechanics and foundation engineering, Moscow, 1973*, 2, 151–158.
15. Nixon, P.J. Floor heave in buildings due to the use of pyritic shales as fill material. *Chem. Ind.*, 1978, Mar., 160–164.
16. Lea F.M. *The chemistry of cement and concrete*, 3rd edn. Edward Arnold, London, 1970.
17. Biczok I. *Concrete corrosion, concrete protection.* Publishing House of the Hungarian Academy of Sciences, Budapest, 1964.
18. Barry D.L. *Material durability in aggressive ground.* Construction Industry Research and Information Association, London, 1983, CIRIA Report 98.
19. Gutt W.H. and Harrison W.H. *Chemical resistance of concrete.* Her Majesty's Stationery Office, London, 1977, Building Research Establishment CP 23/77. *Concrete*, 1977, **11**, No. 5, 35–37.
20. Scottish Malt Distillers Ltd. Personal communication, 1974.
21. Tibbets D.C. *Performance of concrete in sea water. Some examples from Halifax, Nova Scotia. Performance of concrete. Resistance of concrete to sulphate and other environmental conditions.* University of Toronto Press, 1968, 159–180.
22. Flaxman E.W. and Dawes N.J. Developments in materials and design techniques for sewerage systems. *Water Pollut. Control*, 1983, **82**, No. 2, 170–177.
23. Appelton B. Acid test for Middle East drain brains. *New Civ. Engr*, 1976, 19 Feb., 20–23.
24. South African Council for Scientific and Industrial Research. *Corrosion of concrete sewers.* SACSIR, Pretoria, 1969, Series DR 12.
25. Biczok I. *Concrete corrosion, concrete protection.* Publishing House of the Hungarian Academy of Sciences, Budapest, 1964, 316.
26. Lea F.M. *The chemistry of cement and concrete*, 3rd edn. Edward Arnold, London, 1970, 627.
27. Hadderlie E.C. *Monitoring growth rates in wood and rock boring marine bivalves using radiographic techniques. Biodeterioration 5.* Wiley, New York, 1983, 304–318.

CHAPTER 6
1. Building Research Establishment. *Determination of chloride and*

References

cement content in hardened Portland cement concrete. Her Majesty's Stationery Office, London, July 1977, IS 13/77.
2. Building Research Establishment. *Simplified method for the detection and determination of chloride in hardened concrete.* Her Majesty's Stationery Office, London, July 1977, IS 12/77.
3. Treadaway K.W.T. and Page C.L. The durability of steel in concrete. *Building Establishment News*, 1984, **61**, Winter, 4–5.
4. Cavalier P.G. and Vassie P.R. Reinforcement corrosion in a bridge deck. *Proc. Inst. Civ. Engrs*, Part 1, 1981, **70**, Aug., 468–469; discussion, 1982, Aug., 417.
5. Stratfull R.F. The corrosion of steel in a reinforced concrete bridge. *Corrosion*, 1957, **13**, Mar., 173–175.
6. Stratfull R.F. Half-cell potentials and the corrosion of steel in concrete. *High. Res. Rec.*, 1973, No. 423, 1–11.
7. Tremper B., Beaton J.L. and Stratfull R.F. Corrosion of reinforcing steel and repair of concrete in a marine environment. *Highw. Res. Board Bull.*, 1958, No. 182.
8. Lea F.M. *The chemistry of cement and concrete*, 3rd edn. Edward Arnold, London, 1970, 570.
9. Figg J.W. and Bowden S.R. *The analysis of concretes.* Her Majesty's Stationery Office, London, 1970, 67.
10. Building Research Establishment. *Alkali aggregate reactions in concrete.* Her Majesty's Stationery Office, London, February 1982, Digest 258.
11. Terzaghi R.D. Concrete deterioration in a shipway. *J. Am. Concr. Inst.*, 1948, **19**, No. 10, 977–1001.

CHAPTER 7
1. British Standards Institution. *Methods of test for soils for civil engineering purposes.* BSI, London, 1975, BS 1377.
2. Bowley M.J. *Analysis of sulphate-bearing soils in which concrete is to be placed.* Her Majesty's Stationery Office, London, 1979, Building Research Establishment CP 2/79.
3. Building Research Establishment. *Concrete in sulphate-bearing soils and groundwaters.* Her Majesty's Stationery Office, London, 1981, Digest 250.
4. British Standards Institution. *Structural use of concrete. Part 1:Code of practice for design and construction.* Her Majesty's Stationery Office, London, 1985, BS 8110.
5. Building Research Establishment. *Hardcore.* Her Majesty's Stationery Office, London, 1983, Digest 276.
6. British Standards Institution. *British Standard guide to the use of industrial by-products and waste materials in building and civil*

engineering. *Section 3, uses in building*. BSI, London, 1985, BS 6543.
7. British Standards Institution. *Air-cooled blastfurnace slag coarse aggregate for concrete*. BSI, London, 1983, BS 1047.
8. British Standards Institution. *Methods of testing cement. Part 2: Chemical tests*. BSI, London, 1970, BS 4550.
9. British Standards Institution. *Pulverized-fuel ash. Part 1:Specification for pulverized-fuel ash for use as a cementitious component in structural concrete*. BSI, London, 1982, BS 3892.
10. British Standards Institution. *Portland pulverized-fuel ash cement*. BSI, London, 1985, BS 6588.
11. British Standards Institution. *Pozzolanic cement with pulverized-fuel ash as pozzolana*. British Standards Institution, London, 1985, BS 6110.
12. British Standards Institution. *Methods for sampling and testing of mineral aggregates, sands and fillers. Part 4:Methods for determination of chemical properties*. BSI, London, 1976, BS 812.
13. British Standards Institution. *Part 2:Methods for sampling and testing of lightweight aggregates for concrete*. BSI, London, 1973, BS 3681.
14. British Standards Institution. *Part 106:Method for determination of shell content in coarse aggregate*. BSI, London, 1985, BS 812.
15. British Standards Institution. *Aggregates from natural sources for concrete*. BSI, London, 1983, BS 882.
16. British Standards Institution. *Part 119:Method for determination of acid-soluble material in fine aggregate*. BSI, London, 1985, BS 812.
17. American Society for Testing and Materials. *Standard practice for petrographic examination of aggregates for concrete*. ASTM, Philadelphia, 1965, ASTM C295–65, reapproved 1973.
18. American Society for Testing and Materials. *Standard test method for potential reactivity of aggregates (chemical method)*. ASTM, Philadelphia, 1971, ASTM C289–71, reapproved 1976.
19. American Society for Testing and Materials. *Standard test method for potential alkali reactivity of cement – aggregate combinations (mortar-bar method)*. ASTM, Philadelphia, 1971, ASTM C227–71, reapproved 1976.
20. American Society for Testing and Materials. *Standard Specification for concrete aggregates*. ASTM, Philadelphia, 1978, ASTM C33–78.
21. Mielenz R.C. and Greene K.T. Chemical test for the reactivity of aggregates with cement alkalies, chemical processes in cement aggregate reaction. *Proc. Am. Concr. Inst.*, 1947, **44**, 193.
22. Mielenz R.C. and Benton E.J. Evaluation of the quick chemical test for alkali reactivity of concrete aggregates. *High. Res. Board Bull.*, 1958, No. 171.

References

23. Chaiken B. and Halstead W.J. Correlation between chemical and mortar bar tests for potential alkali reactivity of concrete aggregates. *Public Roads*, 1953, June, 177–184.
24. Jones F.E. and Tarleton R.D. *Reactions between aggregates and cement*. Her Majesty's Stationery Office, London, National Building Studies, Research Papers 14, 15, 17, 1952 and 20, 25, 1958.
25. Nixon P.J. and Bollinghaus R. Testing for alkali reactive aggregate in the UK. *Proc. 6th international conference on alkalis in concrete – research and practice, Copenhagen, June 1983*. Danish Concrete Association, Copenhagen, 1983.
26. British Standards Institution. *Methods for testing concrete. Part 6: Analysis of hardened concrete*. BSI, London, 1971, BS 1881.
27. Figg J.W. and Bowden S.R. *The analysis of concretes*. Her Majesty's Stationery Office, London, 1973.

CHAPTER 8
1. Lea F.M. *The chemistry of cement and concrete*, 3rd edn. Edward Arnold, London, 1970.
2. Biczok I. *Concrete corrosion, concrete protection*. Publishing House of the Hungarian Academy of Science, Budapest, 1960.
3. American Concrete Institute. *A guide to the use of waterproofing, damp proofing, protective and decorative barrier systems for concrete*. American Concrete Institute, Detroit, 1979, Report 515, IR–79.
4. Gutt W.H. and Harrison W.H. *Chemical resistance of concrete*. Building Research Establishment CP 23/77, reprinted from *Concrete*, 1977, **11**, No. 5, 35–37.
5. Perkins P.H. *Concrete structures: repair, waterproofing and protection*. Applied Science, London, 1976.
6. Van Arrdt J.P.H. and Fulton F.S. In *Concrete technology*, edited by F.S. Fulton, Portland Cement Institute, Johannesburgh, 1977, 167–193.
7. L'Association Française de Normalisation. *Concretes. Classification of aggressive environments*. L'Association Française de Normalisation, Paris, 1985, French Standard P18–011, May 1985. Building Research Establishment translation, London.
8. Building Research Establishment. *Concrete in sulphate-bearing soils and ground waters*. Her Majesty's Stationery Office, London, 1981, Digest 250.
9. British Standards Institution. *Structural use of concrete. Part 1: Code of practice for design and construction*. BSI, London, 1985, BS 8110.
10. British Standards Institution. *Methods for specifying concrete, including ready-mixed concrete*. BSI, London, 1981, BS 5328.

11. British Standards Institution. *Portland pulverized-fuel ash cement.* BSI, London, 1985, BS 6588.
12. British Standards Institution. *Supersulphated cement.* BSI, London, 1974, BS 4248.
13. British Standards Institution. *Specification for aggregates from natural sources for concrete.* BSI, London, 1983, BS 882.
14. Building Research Establishment. *The durability of steel in concrete. Part 1:Mechanism of protection and corrosion.* Her Majesty's Stationery Office, London, 1982, Digest 263.
15. Cork H.A. Coating treatment for reinforcing steel. *Concrete*, 1977, Jan., 31–32.
16. Burge T.A. *Densified silica-cement coating as an effective corrosion protection.* In *Corrosion of reinforcement in concrete construction*, edited by A.P. Crane. Published for the Society of Chemical Industry by Ellis Horwood, Chichester, 1983, 333–342.
17. Sprung S. Effect of cement and admixtures on the alkali–silica reaction. *Proc. symposium on preventive measures against alkali–silica reactions in concrete, Hamburg, 1973. VD1-Z,* **43**, 69–78.
18. Hobbs D.W. *Influence of mix proportions and cement alkali content upon expansion due to the alkali–silica reaction.* Cement and Concrete Association, Wexham Springs, Slough, 1980, Technical Report 534 (publication 42.534).
19. Building Research Establishment. *Alkali aggregate reactions in concrete.* Her Majesty's Stationery Office, London, 1982, Digest 258.
20. Hawkins M.R. *et al. Alkali–aggregate reactions. Minimising the risk of alkali–silica reaction.* Guidance notes. Report of a working party, 1983. Published for the working party by Cement and Concrete Association, Wexham Springs, Slough.
21. Palmer D. Alkali–aggregate reaction in Great Britain – the present position. *Concrete*, 1981, **15**, No. 3, 24–27.
22. Stanton T.E. The expansion of concrete through reaction between cement and aggregate. *Proc. Am. Soc. Civ. Engrs*, 1940, **66**, Dec., 1781–1811.
23. Sims I. and Poole A.B. Potentially alkali–reactive aggregates from the Middle East. *Concrete*, 1980, May, 27–30.
24. Nixon P.J., Collins R.J. and Rayment P.L. Migration of alkalis through concrete. *Cement and Concr. Res.*, 1979, **9**, No. 4, 417–427.
25. Bartholomew R.F. The protection of concrete piles in aggressive ground conditions: an international appreciation. *Proc. conference on recent developments in the design and construction of piles.* Institution of Civil Engineers, London, 1979, 131–141.

Index

Acids, action on concrete
 carbonic acid, 65, 108
 fulvic acid, 39, 41
 humic acid, 39, 45, 65
 inorganic acids, 69
 naturally occurring acids, 65
 organic acids, 70
 sulphuric acid, 57, 67
Activation of artificial pozzolanas, 15
Activation of blastfurnace slag, 12
Admixtures for concrete, 47
 accelerating, 48
 air-entraining, 51
 retarding, 50
 superplasticising, 51
 waterproofing, 51, 107
 water-reducing, 50, 107
Aggregates for concrete, 21
 alkali reactivity of, 31
 blastfurnace slag, 41
 chlorides in, 27
 clay in, 21
 crushed rock, 23, 29
 dense, 21
 flint, 21, 23, 37, 97, 102
 gravel, marine-dredged, 22, 29, 97
 iron minerals in, 37
 lightweight, 43
 limestones, porosity of, 26
 natural, 21
 organic matter in, 39
 sand, beach, 23
 dune, 23
 marine-dredged, 22, 29, 97
 pit, 23, 29, 39
 shells in, 97
 specifications for, 30, 40, 43
 staining of concrete by, 39, 41
 sulphates in, 22
 washing of, 25, 29
Aluminate phases
 in high alumina cement, 18
 in Portland cements, 6, 8
Aluminium oxide
 in high alumina cement, 18
 in Portland cement, 4, 9
Aluminium silicates, 11, 15
Aluminium stearate, 52
Alkali bicarbonates, 46
Alkali carbonates, 46
Alkali chlorides, 29
Alkali hydroxides, 31, 70
Alkali metals, 31
Alkali metals in Portland cements, 6, 31, 114
Alkali reactions with aggregates, 31, 35, 46, 61, 70, 87, 99
Alkali soaps, 53, 117
Alkali sulphates, 23
Ammonium chloride, 71
Ammonium hydroxide, 71
Ammonium nitrate, 79
Ammonium stearate, 52
Ammonium sulphate, 71
Analysis of blastfurnace slag, 12
Analysis of blastfurnace slag cements, 12
Analysis of sea water, 26, 61

Index

Analytical test methods, 92
Anhydrite, 22
Asphalt coatings, 116

Bacteria
   sulphur-oxidising, action of, 67, 76
   sulphur-reducing, action of, 76
Bauxite in high alumina cement manufacture, 18
Biological attack on concrete, 75
Bituminous coatings, 66, 116
Blastfurnace slag
   as aggregate, 41, 43
   as cement replacement, 13
   cements, 11
   composition of, 12
   crystalline form, 41
   foamed form, 43
   granulated form, 11
      activators for, 12
      ground, use of, 13
      hydration of, 13
      hydraulicity of, 11
   mineralogy of, 41
   tests on, 42
   unsoundness of, 41

Bricks
   as aggregate, 41
   as fill material, 55
   as pozzolanas, 5
Butyl stearate, 50

Calcium aluminate
   conversion of, 18
   in high alumina cement, 18
   in Portland cement, 6, 8
Calcium carbonate, 65, 87
Calcium chloride, 48
Calcium chloroaluminate, 49, 58
Calcium chloroferrite, 49
Calcium hydroxide, 7, 12
Calcium oxide, 4, 11
Calcium silicates, 7, 11
Calcium soaps, 53
Calcium sulphate
   in aggregates, 22
   in cements, 4, 7, 12, 13
   in soils, 53
   in waters, 55, 62
Calcium sulphide in blastfurnace slag, 42
Calcium sulphoaluminate, 7
Capillary water in concrete, determination of, 103
Carbon dioxide
   action on concrete of, 73
   aggressive behaviour of, 65
   in water, 65
Carbonic acid, action on concrete, 65
Cement
   blastfurnace slag, 11
   high alumina, 17
   Portland, 6
   pozzolanic, 15
   raw materials
      for high alumina, 18
      for Portland cement, 6
   replacement materials for, 13, 15
   supersulphated, 13
Cement content of concrete, determination of, 102
Chalcedony, 32
Chalk, 6
Chemical test methods
   for aggregates, 27, 97
   for cements, 96
   for concrete, 85, 102
   for fills, 93
   for pulverised fuel ash, 97
   for soils, 93
   for waters, 93
Chert, 15, 32
Chlorides
   in admixtures, 48
   in aggregates, 27
   corrosion of steel by, 28, 49, 58, 79, 83
Clay
   in aggregates, 21, 36
   bricks as pozzolana, 5
   expanded, as aggregate, 43
   occurrence of sulphates in, 53
   tiles as pozzolana, 5

Index

Concrete
   action of external agents on, 53
   acids, inorganic, 60
   acids, organic, 70
   aggressive carbon dioxide, 65
   alkaline solutions, 69
   ammonium salts, 71
   chlorides, 57
   creosote, 73
   fats, 72
   fruit juices, 72
   gases, 73
   glycerol, 72
   oils, 72
   phenols, 73
   salts, 69
   sea water, 61
   soft waters, 66
   sugars, 71
   sulphates, 53
   analysis of, 102
   corrosion of steel in, 28, 46, 49, 57, 79, 83
   cracking of, 21, 22, 28, 32, 35, 56, 79, 87
   definition of, 1, 2
   durability of, 1
   influence on
      aggregate grading, 2
      aggregate particle shape, 2
      curing, 3
      water/cement ratio, 2
   leaching of lime from, 7, 64, 65, 71, 90, 91
   microscopical examination of, 80
   mixing water for, 45
   protective coatings for, 66, 111, 115
   staining of, 37, 39, 79
Creosote, action on concrete of, 73
Crystobalite, 32

Dactite, 32
Dams
   acid waters in, 66
   soft waters in, 66
Diatomaceous earth, 15
Dicalcium silicate
   in blastfurnace slag, 41
   in Portland cement, 6
Dolomite, alkali reactions of, 35

Epoxy resins, 110
Ettringite, 7, 14, 22, 54, 63, 80, 91
Examination of concrete failures, 78
   chemical tests for, 81, 84, 87
   microscopical, 80, 88
   visual, 80, 83, 87, 96
   X-ray techniques, 80
Expanded materials as aggregates, 43
   blastfurnace slag, 43
   clay, 43
   perlite, 43
   shale, 43
Expansive chemical reactions
   alkali–carbonate, 35
   alkali–silica, 32
   alkali–silicate, 35
   magnesium oxide, 10, 13
   pyrite, 68
   sulphates, 24, 81

Failure of concrete, 78
Fats, effect on concrete, 72
Ferric oxide
   in high alumina cement, 18
   in Portland cement, 6
Ferric sulphate, 67
Ferrite phases in Portland cements, 6, 9
Field test methods, 27, 31, 84
Fills, 55, 81
Flint aggregate, alkali–silica reactions of, 33
Fly ash as cement replacement
   as pozzolana, 15
   sintered as aggregate, 43
Foamed blastfurnace slag aggregate, 43
Fulvic acid, action on concrete, 39, 41

Galvanised steel reinforcement, 110
Gases, action on concrete, 73
Gels, alkali–silicate, 33, 87

133

Index

Glass
  blastfurnace slag, 11
  volcanic, 5
Glycerol, effect on concrete, 72
Granite, 21, 38
Granulated blastfurnace slag, 11
  ground as cement replacement, 13
Gravel aggregates, marine-dredged, 22, 29, 97
Gypsum in aggregate, 22
  in cements, 6
  in soils, 53

High alumina cement, 17
  alkaline hydrolysis of, 19
  composition of, 18
  conversion of, 18
  hydration of, 18
  use in refractory concrete, 20
Humic acid, action on concrete, 39, 45, 65
Hydraulic cements, 4
Hydraulic lime, 5
Hydrogen sulphide
  action on concrete of, 74
  in sewage, 76
Hydroxycarboxylic acids, 50

Igneous rocks as aggregates, 23
Industrial chemicals, action on concrete, 68
Inorganic acids, action on concrete, 69
Insoluble matter in cement, 10
Iron compounds
  in aggregates, 37
  in biotite, 38
  in cements, 6, 11, 18
Iron pyrite
  oxidation of, 37
  reactive forms of, 37
  staining by, 37
  sulphuric acid from, 37, 67
Iron unsoundness in blastfurnace slags, 41

Laboratory test methods, 27, 80, 92
Larnite in blastfurnace slags, 41

Leaching of concrete, 7, 17, 63, 64, 65, 71, 90, 91
Lightweight aggregates, 43
  chemical requirements for, 43
  clinker, 44
  exfoliated vermiculite, 43
  expanded clay, 43
  perlite, 43
  shale, 43
  foamed blastfurnace slag, 43
  furnace bottom ash, 44
  pelletised granulated blastfurnace slag, 43
  pumice, 43
  sintered fly ash, 43
Lignosulphonates, 50
Lime, 4
Lime saturation factor, 9
Limestone, 18, 26, 32, 77
Limits of constituents
  in aggregates, 23
    chloride, 27
    loss on ignition, 44
    sulphate, 23
  in cements, 8, 14, 16, 18
    alkalis, 32
    insoluble matter, 10
    loss on ignition, 10
    magnesium oxide, 10
    sulphate, 10
  in concrete
    chloride, 30
    sulphate, 24
Linseed oil, 117

Magnesium oxide
  in blastfurnace slag, 13
  in Portland cement, 9
Magnesium silicates, 10, 11
Magnesium silicofluoride, 117
Magnesium sulphate, 22, 53
Magnetite, 38
Marl, 6
Metallic coatings for steel, 110
Mica
  as filler for paints, 117
  staining by iron minerals in, 38
Microscopical examinations of concrete, 80

Index

Mineral oils, action on concrete, 72
Mixing water for concrete, 45
Molecular volumes, 18, 22

Naturally occurring acids, action on concrete, 65
Non-destructive test methods, 85

Oils, effect on concrete, 72
Opal, 15, 32
Organic acids, action on concrete, 70
Organic matter
 in aggregates, 39
 in waters, 45
 staining by, 39, 41
 tests for, 40, 41

Paints for concrete, 117
Pat test for alkali reactivity of aggregates, 100
Peat, 39
Periclase, 10, 13
Perlite, 43
pH value, 9, 14, 19, 46, 69, 93, 107
Phenols, effect on concrete, 73
Phyllites, 35
Pitch coatings, 73, 113, 116
Polymers in concrete, 52, 108
Portland cements
 aluminate phases in, 6, 8
 blastfurnace slag, 11
 composition of, 6
 ferrite phases in, 6, 8
 ordinary, 6
 pozzolanic, 15
 silicate phases in, 6
 specifications for, 8, 10
 sulphate-resisting, 8
Potassium hydroxide, 31, 70
Potassium nitrate, 79
Potassium oxide, 31
Potassium sulphate, 23, 57
Pozzolanas, 15
 activity of, 15
 alkalis in, 114
 artificial, 15
 composition of, 15
 fly ash as, 15

 natural, 5, 15
 tests on, 16, 96
Pozzolanic cements, chemical resistance of, 17, 107, 117
Protection of concrete against chemical attack, 105
Protective coatings
 for concrete, 107, 116
 for steel, 110
Pulverised fuel ash
 as aggregate, 43
 as cement replacement, 108, 114
 composition of, 16
 in pozzolanic cement, 15
Pumice, 43
Pyrite, 37, 59, 67, 79, 94
Pyrrhotite, 37

Quartz, 32
Quartzite, 32

Raw materials for cements
 blastfurnace slag, 11
 high alumina, 18
 Portland, 6
 pozzolanic, 15
Refractory concrete, 20
Reinforcement
 corrosion of, 28, 46, 49, 57, 79, 83
 corrosion inhibitors for, 112
 protective coatings for, 110
Resins, synthetic, 52, 66, 110
Resistance of cements to chemical attack
 by acids, 17, 19
 by sea water, 17
 by soft waters, 17, 19
 by sulphates, 17, 19
Retarding admixtures for concrete, 50

Sampling concrete, 79
Sand as aggregate for concrete, 22, 30
Sandstone, 21
Saponification of oils, 72
Sea water
 action on concrete, 61

composition of, 62
concrete attacked by, 81
Sewage, 75
Shales, 15, 32, 43, 55, 68
Silage, 70
Silica, reactive forms of, 32
Silicate minerals, 4, 15, 35, 43
Silicate phases in Portland cements, 6, 8
Silicofluorides, 117
Soaps as waterproofers, 53, 117
Sodium calcium silicate gels, 33, 87
Sodium chloride, 29
Sodium hydroxide, 31, 74
Sodium oxide, 31
Sodium stearate, 51
Sodium sulphate, 22, 53
Soils
  acid, 65, 67
  organic, 65
  sulphates in, 53
  tests on, 93
Soluble salts
  in aggregates, 27
  in soils, 53
Specifications
  for admixtures, 48
  for aggregates, 30, 41, 43
  for cements, 9, 12, 16, 18
  for chemical analysis, 92
  for pulverised fuel ash, 16
Staining of concrete, 37, 39, 79
Steel, corrosion of, 28, 46, 49, 57, 79, 83
Sugars, action on concrete, 71
Sulphates
  action on concrete, 53
  occurrence
    in soils, 53
    in waters, 53, 62
Sulphides in blastfurnace slag, 42
Sulphuric acid
  action on concrete, 69, 75
  produced by bacterial action, 69, 75
  produced by oxidation of pyrite, 67

Superplasticising admixtures for concrete, 51
Surface coatings for concrete, 66, 111, 115

Tar coatings, 73, 113, 118
Tetracalcium aluminoferrite
  action of chlorides on, 28, 49
  action of sulphates on, 54
  in Portland cement, 6, 8
Thermal methods of analysis, 89
Tiles, ground, as pozzolana, 5
Tricalcium aluminate
  action of chlorides on, 28, 49, 58
  action of sulphates on, 53, 62
  effect of gypsum on setting of
  in Portland cement, 6, 7
Tricalcium silicate in Portland cement, 6, 7
Tricalcium sulphoaluminates, 7
  see also ettringite

Vegetable oils, action on concrete, 72
Vermiculite, 43
Volume increase in chemical reactions, 10, 25, 28, 68

Water, for mixing concrete, 45
Water/cement ratio, determination of, 103
Waterproofers
  integral, 51
  surface, 66, 106, 113, 116, 117
Water-reducing admixtures, 51
Waters, action on concrete
  acidic waters, 50, 70
  sea waters, 61
  soft waters, 66
  sulphate waters, 53, 61

X-ray diffraction, analysis by, 6, 80, 93

Zinc coating on steel, 110
Zinc silicofluoride, 117